D0909326

PROJECTIVE GEOMETRY
AND ALGEBRAIC STRUCTURES

PROJECTIVE GEOMETRY AND ALGEBRAIC STRUCTURES

R. J. Mihalek
University of Wisconsin-Milwaukee

Academic Press
New York and London

ACADEMIC PRESS, INC.
111 Fifth Avenue, New York, New York 10003

United Kingdom Edition published by
ACADEMIC PRESS, INC. (LONDON) LTD.
24/28 Oval Road, London NW1 7DD

LIBRARY OF CONGRESS CATALOG CARD NUMBER: 78-159621

AMS (MOS) 1970 Subject Classifications: 50-01, 30-01

PRINTED IN THE UNITED STATES OF AMERICA

Cover design by Jorge Hernandez

CONTENTS

Chapter 5 THEOREMS OF DESARGUES AND PAPPUS

Chapter 6 COORDINATIZATION

Chapter 7 PROJECTIVITIES

Chapter 8 HARMONIC QUADRUPLES

Chapter 9 THE REAL PROJECTIVE PLANE

Chapter 10 PROJECTIVE SPACES—Part 1

Chapter 11 PROJECTIVE SPACES—Part 2

Appendix A HILBERT'S AXIOMS FOR A EUCLIDEAN PLANE

PREFACE

This book is intended as a text for an undergraduate course in projective geometry ordinarily taught at the junior–senior level. The study of projective geometry is particularly suitable for introducing students to the so-called mathematical method. We realize this in the present setting through the consideration of several axiomatic systems, the construction of numerous examples illustrating these systems, and a study of interrelations among these systems.

Affine planes are studied first in the interest of staying close to the students' probable past experience in euclidean geometry. Projective planes are then considered. Next the connection between affine and projective planes is carefully established. Here, the concepts of isomorphism and imbedding are presented, concepts fundamental in the study of modern mathematics.

The main theme of this work introduces the student to the relationship between geometry and algebra. Indeed, this may be considered as an introduction to abstract algebra. For some students, algebraic concepts appear less "abstract" when studied first in a geometric setting. This study provides then another reason for abstracting in the algebraic context. The study of this relationship is developed to include the real projective plane and the system of real numbers.

Finally, an introduction to the study of projective spaces is taken up.

Euclidean geometry is used in motivating much of this study. A modern abstract point of view is adopted, giving historical considerations, to a large degree, a secondary role. At the outset, the details of the motivation are kept to a minimum. This is done to expedite the beginning of the theory. Then as the theory is developed and the need arises, the motivational discussions are resumed. Hence for the most part, the motivational discussions immediately precede the introduction of new concepts.

Some fundamentals from set theory are used here. The usage, however, is from a very intuitive point of view. For an elaboration of these set-theoretic ideas and for further discussions of the mathematical method, references [6], [9], [13], and [20] are suggested reading. (Numerals in brackets refer to the correspondingly numbered items in the list of references.) Reference [10] is suggested for a more advanced point of view concerning the set-theoretic ideas.

Each chapter and section is prerequisite to the units that follow it with the following exceptions:

(a) Sections 3.4, 6.6, and 7.3 are not necessary for subsequent material;
(b) Chapters 10 and 11 can be studied after Section 5.2.

Experience has shown that the material can be covered in from one to one and one-half semesters (meeting three hours per week). References [2], [3], [4], [5], [6], [8], [14], [15], and [18] are suggested for further reading and as sources for additional references.

ACKNOWLEDGMENTS

The author wishes to acknowledge his indebtedness to colleagues and students for helpful suggestions and encouragement during the writing of this book. In particular, Professor Karl Menger first introduced the author to projective geometry and made lasting impressions which influence the spirit of this book.

INTRODUCTION CHAPTER 1

We shall, ultimately, study projective geometry as an abstract theory, self-contained and independent. Nevertheless, we should be remiss not to take advantage, initially, of our earlier experiences in geometry. Such experiences are generally realized in *euclidean geometry*, more specifically, in the high school mathematics traditionally encountered during the sophomore year. We are not interested in the study of euclidean geometry per se, but merely in its use to motivate and to illustrate the current theory. A less precise formulation of euclidean geometry will generally suit our purpose. However, for a thorough treatment of the subject, were the need to arise, we could refer to "The Foundations of Geometry" [12], the notable work of David Hilbert (1862–1943).

We shall assume, therefore, a familiarity with euclidean geometry, something close to the high school experience. Presumably, then, we share a common frame of reference as we begin our introductory discussion.

1.1. EUCLIDEAN PLANES

In his theory of euclidean planes, Hilbert uses *point*, *line*, *incidence*, *betweenness* and *congruence* (of *line segments* and *angles*) as primitive (or undefined) concepts. (See Appendix A for the foundation of Hilbert's theory. This is based on a later version; the first appeared in 1899.) The points and lines are the "objects" of the theory; incidence is a "relation" between points and lines; betweenness is a "relation" on triples of points; and congruence is a "relation" on pairs of line segments and on pairs of angles. He also states 15 axioms that the primitive concepts are assumed to satisfy. Then additional concepts are defined and additional statements (theorems) are stated and proved. This constitutes the theory.

We have, then, a collection of concepts (both primitive and defined) and a body of statements (both assumed and proved) in the euclidean theory. Now we can look for an alternate development of the theory by starting with different primitive concepts (selected from our collection of concepts) or different axioms (selected from our body of statements). Then the remaining concepts would be recovered through definition and the remaining statements through proof. Such an alternate development of a theory might be undertaken to understand better the significance of certain concepts and their properties, to connect the theory with another apparently unrelated theory, to make the theory more readily applicable to other situations, or merely to satisfy a personal curiosity. In our study of projective geometry, we shall be undertaking, in a sense, such an alternate development of euclidean geometry. Let us make the following observations in this vein.

The concept of *projection* is of interest and importance in euclidean geometry. Projections can be used to depict how an object casts a shadow (see Figure 1.1). Also, projections are used in describing the optics of a camera or a projector.

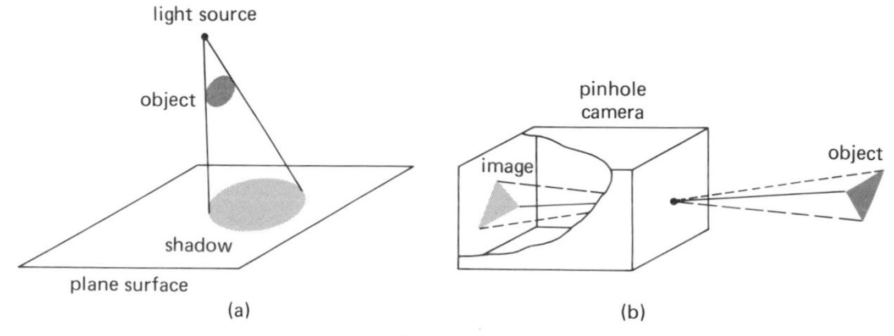

Figure 1.1

Let $\mathscr{A}$ and $\mathscr{A}'$ be planes in euclidean space and O a point not on the planes (see Figure 1.2 where $\mathscr{A}$ and $\mathscr{A}'$ are parallel or Figure 1.3 where $\mathscr{A}$ and $\mathscr{A}'$ intersect). Then under the projection from $\mathscr{A}$ to $\mathscr{A}'$ with *center* O, point P' on $\mathscr{A}'$ is the correspondent of point P on $\mathscr{A}$ if O,

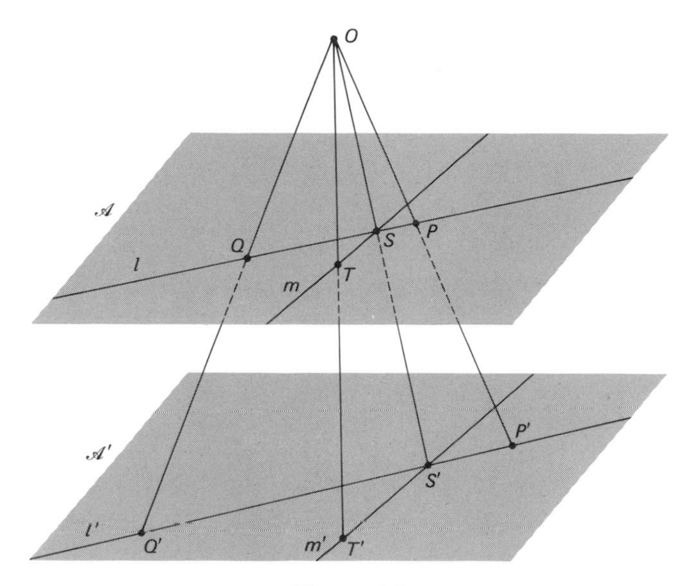

Figure 1.2

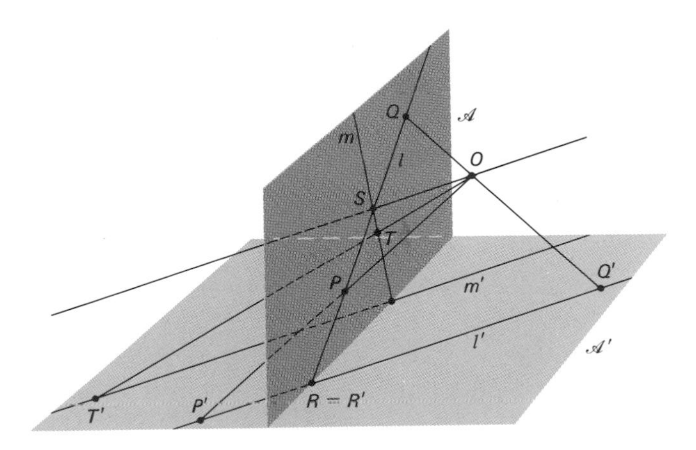

Figure 1.3

P, and P' are collinear. Also line l' on $\mathscr{A}'$ is the correspondent of line l on $\mathscr{A}$ if at least two points on l' are correspondents of points on l.

Now we ask if any of the euclidean concepts are "invariant" under projections. This means, for points, that each point on $\mathscr{A}$ has a correspondent on $\mathscr{A}'$ and each point on $\mathscr{A}'$ is the correspondent of some point on $\mathscr{A}$. A similar meaning is ascribed for the "invariance" of lines. "Invariance of incidence" means that a point and line on $\mathscr{A}$ are incident if and only if their correspondents (when existent) on $\mathscr{A}'$ are incident. For parallelism to be invariant, we should want lines on $\mathscr{A}$ to be parallel if and only if their correspondents (when existent) on $\mathscr{A}'$ are parallel. The meaning of "invariant" for the other euclidean concepts is formulated similarly.

In case the planes $\mathscr{A}$ and $\mathscr{A}'$ are parallel (as in Figure 1.2), the situation is readily described. The concepts of point, line, incidence, betweenness, and parallelism, to name a few, are invariant while the congruence of lines segments is not. (Why?) This suggests studying concepts and statements that depend only on these "invariants" or, for that matter, an alternate development of the theory of euclidean planes using new primitive concepts selected from among these invariants and new axioms selected from among the statements involving only these concepts. Indeed, we begin such a study in Chapter 2 with *affine planes*, employing "point," "line," and "incidence" as primitive concepts. However, in order to continue the study, we must bring some additional considerations into play, in particular, some that arise when the planes $\mathscr{A}$ and $\mathscr{A}'$ are not parallel.

Let the planes $\mathscr{A}$ and $\mathscr{A}'$ intersect as in Figure 1.3. Then in the figure, points P, Q, R, T, on $\mathscr{A}$ have correspondents P', Q', R', T', respectively, on $\mathscr{A}'$. The point S on $\mathscr{A}$ is such that the line on O and S is parallel to $\mathscr{A}'$. Hence S has no correspondent on $\mathscr{A}'$. Thus the concept of point is not invariant under this projection. A similar situation holds for lines. (What line on $\mathscr{A}$ has no correspondent on $\mathscr{A}'$ and what line on $\mathscr{A}'$ is not the correspondent of any line on $\mathscr{A}$?) Incidence is invariant. (Why?) Parallelism is not; consider lines l and m on $\mathscr{A}$ in the figure. In addition, betweenness is not invariant. Consider, in the figure, P between Q and R while P' is not between Q' and R'. Congruence fails to be invariant for this projection also.

When comparing the two projections considered, it appears that the loss of invariance for some of the concepts relative to the latter projection is related to parallelism. Indeed, in Figure 1.3, point S has no correspondent on $\mathscr{A}'$ because the line on O and S is parallel to $\mathscr{A}'$. Early geometers recognized this situation and circumvented it by devising a scheme through which they introduced additional objects (called *ideal points*, *ideal lines*, and an *ideal plane*) to eliminate parallelism. In other words, they enlarged euclidean space to a geometry without parallelism. As might be expected in this enlarged geometry, many of the analogs of euclidean concepts and statements undergo considerable alteration. For example, any two "lines" on a "plane" will "intersect." Nevertheless, by design, the concepts in this enlarged geometry that are the analogs of euclidean point, line, incidence, and betweenness are invariant under projections. We shall undertake, then, an alternate development of the theory of this enlarged geometry using only these invariant concepts. We begin this development in Chapter 3 with the study of *projective planes*, employing "point," "line," and "incidence" as primitive concepts. We do not introduce the analog of betweenness, called *separation*, until Chapter 9. A considerable bulk of the "planar" theory of this enlarged geometry can be developed before separation is needed. We continue the alternate development of the enlarged geometry with the study of *projective spaces* in Chapters 10 and 11.

We choose not to describe, at this time, the above "enlargement" scheme. Consequently, we cannot know the details of the properties of the enlarged geometry. However, this circumstance presents no deterrent to our study. We already have an idea of what we need for affine and projective planes. In Chapter 4, we describe the relation between these two concepts, which is, in effect, the "enlargement" scheme for "planes." Thereafter, we introduce and discuss properties of the enlarged geometry as needed.

We are able to use the concept of invariance (under projections) to classify geometric properties as *descriptive* or *metric*. Properties involving only the invariant concepts are said to be descriptive. They are the properties concerned with positional relationships of geometric objects such as incidence and betweenness. The metric properties involve the noninvariant concepts. Such properties include the measurement of distances, angles, and areas.

Studies of descriptive properties in geometry go back to antiquity. The French mathematician Gérard Desargues (1593–1662) published a treatise in 1639 on the conic sections using descriptive properties. The significance of his work was not generally recognized until nearly two centuries later. The consideration of descriptive properties was introduced again into the study of geometry by Jean Victor Poncelet (1788–1867) early in the last century. He published his great work on projective geometry in Paris in 1822. This is generally accepted as the beginning of the modern period in the development of projective geometry.

Throughout the last century many took up the study of projective geometry, including men such as Gergonne, Brianchon, Chasles, Plücker, Steiner, and Von Staudt. This study, however, was still largely dependent on euclidean geometry; attention was still being focused on descriptive properties of the enlarged geometry. At the turn of the century through the efforts of Felix Klein (1849–1925), Oswald Veblen (1880–1960), Hilbert, and others, projective geometry was developed as an abstract theory. Projective geometry was, then, based on a set of axioms, not dependent on euclidean geometry. In Chapter 3, we begin our study of projective geometry from this point of view. For more of the details of this historic development, references [6], [9], [15], and [18] are suggested reading.

EXERCISES

1.1. Formulate a definition of "invariance" for (a) betweenness, (b) congruence of line segments, (c) congruence of angles. (What is the correspondent of a line segment?)

1.2. In Figure 1.3, find a line on $\mathscr{A}$ with no correspondent on $\mathscr{A}'$ and a line on $\mathscr{A}'$ which is not the correspondent of a line on $\mathscr{A}$.

1.3. Illustrate in Figure 1.3 why congruence (of both line segments and angles) fails to be invariant.

1.2. INCIDENCE BASES

As indicated in the last section, our study will employ, initially, the primitive concepts "point," "line," and "incidence." In other words, when we state axioms for our theory, these words will be undefined. More explicitly, our study will include a collection or set of objects called "points" and a collection or set of objects called "lines." Let us designate these sets by the symbols $\mathscr{P}$ and $\mathscr{L}$, respectively. It will be desirable that we not confuse points with lines. Hence we should like that our sets $\mathscr{P}$ and $\mathscr{L}$ be *disjoint*, that is, have no objects or elements in common. We begin our study, then, with two disjoint sets $\mathscr{P}$ and $\mathscr{L}$ of elements.

Additionally, we want to consider certain points in the "incidence" relation with certain lines, that is, certain points are to be matched or paired with certain lines. The notion of incidence, then, involves a pairing of elements from $\mathscr{P}$ with elements from $\mathscr{L}$. More specifically, we like to speak of ordered pairing. To realize this, we shall use the notion of an *ordered pair* (a, b), where a and b are the elements of the pair with a the first and b the second. Then in our instance, we shall be considering certain ordered pairs (P, l) with P an element of $\mathscr{P}$ and l an element of $\mathscr{L}$. The set of all ordered pairs (P, l) that can be formed with P ranging over the elements of $\mathscr{P}$ and l ranging over the elements of $\mathscr{L}$ is called the *cartesian product* of $\mathscr{P}$ and $\mathscr{L}$ and designated by $\mathscr{P} \times \mathscr{L}$. Hence the study of incidence involves certain ordered pairs from $\mathscr{P} \times \mathscr{L}$, more explicitly, a certain *subset* of $\mathscr{P} \times \mathscr{L}$, that is, a set consisting of elements of $\mathscr{P} \times \mathscr{L}$. Let us designate this subset of $\mathscr{P} \times \mathscr{L}$ by the symbol $\circ$. We say that $\circ$ is a *relation on* $\mathscr{P} \times \mathscr{L}$. We are, now, in a position to summarize the above in an abstract formulation of the systems that we want to study. This is merely a precise way of stating what our primitive concepts are to be.

The system $(\mathscr{P}, \mathscr{L}, \circ)$ will be called an *incidence basis* if $\mathscr{P}$ and $\mathscr{L}$ are disjoint sets and $\circ$ is a relation on $\mathscr{P} \times \mathscr{L}$. Let $(\mathscr{P}, \mathscr{L}, \circ)$ be an incidence basis. The elements of $\mathscr{P}$ will be called *points*, the elements of $\mathscr{L}$ *lines* and $\circ$ the *incidence relation*. If the ordered pair (P, l) is an element of $\circ$, we shall write $P \circ l$, which is to be read "P is on l." In case (P, l) is not an element of $\circ$ for P a point and l a line, we shall write $P \not\circ l$ and say that "P is not on l."

It remains, now, to state axioms for a given incidence basis $(\mathscr{P}, \mathscr{L}, \circ)$ to complete the foundation of our theory. We shall be studying more than one theory, so that we cannot state a single set of axioms at this time. We shall, instead, state the appropriate axioms at the beginning of the study of each particular theory. There is one axiom, however, that will be common to all of the theories. It is the following.

AXIOM. If P and Q are elements in $\mathscr{P}$ such that $P \neq Q$, then there exists one and only one element l in $\mathscr{L}$ such that $P \circ l$ and $Q \circ l$.

In light of the above definitions and discussion, this statement can be rewritten in the following intuitive form.

AXIOM. If P and Q are distinct points, then there exists one and only one line l such that P is on l and Q is on l.

In the development of our theories, we shall make repeated applications of this axiom, which will necessitate, then, the introduction of a new symbol for the line l corresponding to each pair of distinct points P and Q in a particular discussion. This involves an extra step at each application. Further, in a particular discussion, there may be several applications of the axiom, thereby necessitating the introduction of several different symbols for the lines l. This can be awkward if the number is large. We shall, therefore, introduce a standard symbol for the line, which incorporates symbols for the given points. In other words, for P and Q distinct points, the symbol $P \vee Q$ will be used for the unique line l of the Axiom. Hence $P \vee Q$ is a line. More precisely, for P and Q elements of $\mathscr{P}$ such that $P \neq Q$, $P \vee Q$ is the unique element of $\mathscr{L}$ such that $P \circ P \vee Q$ and $Q \circ P \vee Q$ as guaranteed by the axiom. Moreover, we shall call $P \vee Q$ the *join of P and Q*. Finally, we shall find it convenient to define $P \vee P = P$ for P a point.

Any theory with an incidence basis and the above axiom in its foundation will be called an *incidence geometry*. It is our objective in this work to study certain incidence geometries.

1.3. SET THEORY

We employed some basic notions from set theory when formulating a precise statement of the primitive concepts in the last section. This appeal to set theory facilitated the precision. We should like to retain this level of precision. Consequently, we shall need a few additional set-theoretic notions, many of which are included in studies of modern school mathematics. We introduce (or possibly, review) the concepts to be used and select a notation for them in the discussion below. Some of the statements are merely explanatory; some are definitions, and some are properties (stated without proof). Mastery of the list of statements is not a prerequisite for beginning the next chapter. Instead, the study of any particular concept can be delayed until its first appearance in the text. In fact, this is a suggested procedure for anyone who is not especially familiar with the set-theoretic notions.

The notions of *element*, *set*, and "is an element of" will be considered primitive. We shall write $x \in S$ when x is an element of the set S and $x \notin S$ when x is not. For S and T sets, we shall call S a *subset* of T and write $S \subset T$ (also $T \supset S$) if $x \in T$ whenever $x \in S$. Sets S and T are equal if and only if $S \subset T$ and $T \subset S$. The symbol $\{a_1, \ldots, a_n\}$ will be used for the finite set consisting of the elements $a_1, \ldots, a_n$ and $\{x \in S$: x satisfies $\cdots\}$ for the subset of S consisting of the elements x satisfying $\cdots$. The latter set will also be written $\{x : x$ satisfies $\cdots\}$. We shall designate the *empty set* by $\varnothing$.

Let S and T be sets. Define:

(a) the (*set-theoretic*) *union of S and T*, written $S \cup T$, to be

$$\{x : x \in S \quad \text{or} \quad x \in T\};$$

(b) the (*set-theoretic*) *intersection of S and T*, written $S \cap T$, to be

$$\{x : x \in S \quad \text{and} \quad x \in T\};$$

(c) the *relative complement of S in T*, written $T - S$, to be

$$\{x \in T : x \notin S\}.$$

We shall say that S and T are *disjoint* if $S \cap T = \varnothing$.

The notion of *ordered pair* will also be primitive in our considerations. The notation (a, b) will be used to designate the ordered pair consisting of a and b with a the first element and b the second. In addition, $(a, b) = (c, d)$ will mean $a = c$ and $b = d$. Let S and T be sets. Define the *cartesian product of S and T*, written $S \times T$, to be $\{(x, y) : x \in S$ and $y \in T\}$. A set $R \subset S \times T$ will be called a *relation on $S \times T$*. For R a relation on

$S \times T$, define the *domain of* R to be $\{x \in S : (x, y) \in R$ for some $y \in T\}$ and the *range of* R to be $\{y \in T : (x, y) \in R$ for some $x \in S\}$.

Let S and T be sets and $R \subset S \times T$. We shall say that R is a *function from S to T* if (x, y_1), $(x, y_2) \in R$ implies $y_1 = y_2$. Moreover, if the domain of R equals S, we shall say that R is *on* S to T.

Let S and T be sets and F a function from S to T. For x an element of the domain of F, define the symbol Fx to be that element $y \in T$ such that $(x, y) \in F$. Hence $(x, Fx) \in F$ for x an element of the domain of F. We shall say that F is *one-to-one* if $x_1 = x_2$ whenever $Fx_1 = Fx_2$ for x_1, x_2 elements of the domain of F. When F is one-to-one, the set $\{(y, x) : (x, y) \in F\}$ is a function from T to S, which we call the *inverse of* F and designate by F^{-1}. Then

$$F^{-1}(Fx) = x \qquad \text{for} \quad x \text{ an element of the domain of } F$$

and

$$F(F^{-1}x) = x \qquad \text{for} \quad x \text{ an element of the domain of } F^{-1}$$

whenever F is one-to-one. The function F will be called a *one-to-one correspondence between S and T* if domain of $F = S$, range of $F = T$, and F is one-to-one. If F is a one-to-one correspondence between S and T, then F^{-1} is a one-to-one correspondence between T and S.

Let S, T, U be sets, F a function from S to T and G a function from T to U. The *composite of G and F*, written GF, is a function from S to U defined by $(GF)x = G(Fx)$ for x an element of the domain of F such that Fx is an element of the domain of G. If F is a one-to-one correspondence between S and T and G is a one-to-one correspondence between T and U, then GF is a one-to-one correspondence between S and U.

Let S be a set and n be a positive integer. A function with domain $\{1, 2, \ldots, n\}$ and range contained in S will be called an *n-tuple in S*. The n-tuple in S consisting of the ordered pairs (i, a_i) for $i = 1, \ldots, n$ will be designated by $(a_1, \ldots, a_n)$. It is common to refer to a 3-tuple as an *ordered triple*, a 4-tuple as an *ordered quadruple*, and so on. A function with domain consisting of the positive integers and range contained in S will be called a *sequence in S*. The sequence in S consisting of the ordered pairs (i, a_i) for i a positive integer will be designated by (a_i). The set $F = \{(x, x) : x \in S\}$ will be called the *identity function in S*, and

$Fx = x$ for $x \in S$. A function G from $S \times S$ to S will be called a *binary operation*. For G a binary operation from $S \times S$ to S and (x, y) an element of the domain of G, define $x \; G \; y = G(x, y)$.

Finally, we consider two additional concepts which are variations on cartesian product and relation. Let S_1, S_2, S_3, and S_4 be sets. Define the *cartesian product of* S_1, S_2, S_3, *and* S_4, written $S_1 \times S_2 \times S_3 \times S_4$, to be

$$\{(x_1, x_2, x_3, x_4) : x_i \in S_i \quad \text{for} \quad i = 1, 2, 3, 4\}.$$

A set $R \subset S_1 \times S_2 \times S_3 \times S_4$ will be called a *relation on*

$$S_1 \times S_2 \times S_3 \times S_4 .$$

AFFINE PLANES

CHAPTER 2

We begin our study by considering those incidence bases that are defined below to be affine planes. The notion of parallelism is employed in the description of these geometries. Affine planes are studied first in the interest of remaining closer to what might be a more intuitive conception of geometry. We shall establish properties of affine planes that will enable us later to relate these geometries to projective planes. This will become the link between the reader's (probable) past experience in euclidean geometry and the geometries that are the objects of study in this book.

2.1. AXIOMS FOR AN AFFINE PLANE

Let the system $(\mathscr{P}, \mathscr{L}, \circ)$ be an incidence basis as described in the last chapter. We give two definitions that are convenient for the statement of the axioms.

The points P, Q, R are said to be *collinear* if there exists a line l such that $P \circ l$, $Q \circ l$, and $R \circ l$. For distinct lines l and m, we shall say l is *parallel* to m, written $l \parallel m$, if $P \circ l$ and $P \circ m$ for no point P. We shall write $l \nparallel m$ if l is not parallel to m. A line never will be considered to be parallel to itself. It follows readily that parallelism is symmetric, that is, if, for distinct lines l and m, $l \parallel m$, then $m \parallel l$.

An incidence basis $(\mathscr{P}, \mathscr{L}, \circ)$ is said to be an *affine plane* if the following are satisfied:

AXIOM A_1. For distinct points P and Q, there exists one and only one line l such that $P \circ l$ and $Q \circ l$.

AXIOM A_2. For P a point and l a line such that $P \not\circ l$, there exists one and only one line m such that $P \circ m$ and $m \parallel l$.

AXIOM A_3. There exist three noncollinear points.

Throughout the remainder of this section, assume that $(\mathscr{P}, \mathscr{L}, \circ)$ is an affine plane.

The unique line l that exists, by virtue of Axiom A_1, for distinct points P and Q will be called the *join* of P and Q and will be written $P \vee Q$. We define $P \vee P = P$. Note that $Q \vee P = P \vee Q$. A similar uniqueness holds for a point that is on a pair of distinct lines. This is stated in the first theorem.

THEOREM 2.1. If l, m are distinct lines and P is a point such that $P \circ l$ and $P \circ m$, then P is the only point with this property.

This unique point will be called the *intersection* of l and m and written $l \wedge m$. We define $l \wedge l = l$. Note that $l \wedge m = m \wedge l$.

PROOF OF THEOREM 2.1. Let l and m be distinct lines with $P \circ l$ and $P \circ m$ for some point P. Now let Q be a point with the same property, that is, let $Q \in \mathscr{P}$ such that $Q \circ l$ and $Q \circ m$. The object is to show that $Q = P$. Suppose $Q \neq P$. Then, from Axiom A_1, we have the existence and uniqueness of line $P \vee Q$. Now $l = P \vee Q$ since $P \circ l$ and $Q \circ l$, and $m = P \vee Q$ since $P \circ m$ and $Q \circ m$. Thus $l = m$, contrary to hypothesis. Hence the supposition that $Q \neq P$ must be false.

THEOREM 2.2. If l, m, n are distinct lines such that $m \parallel n$ and for some point P, $P \circ l$ and $P \circ m$, then there exists a point Q such that $Q \circ l$ and $Q \circ n$.

PROOF. Let l, m, n be distinct lines with $m \parallel n$ and let P be a point such that $P \circ l$ and $P \circ m$ (see Figure 2.1). Suppose that $Q \circ l$ and

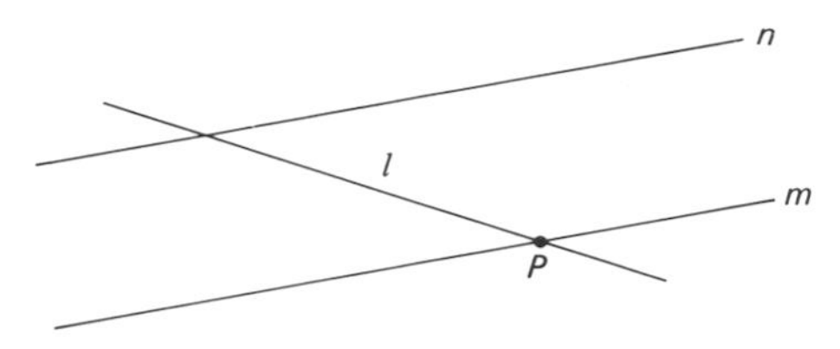

Figure 2.1

$Q \circ n$ for no point Q, that is, $l \parallel n$. Then we have $P \circ l$ with $l \parallel n$ and $P \circ m$ with $m \parallel n$. Hence, by the uniqueness part of Axiom A_2, we have $l = m$ contrary to our hypothesis. Thus $Q \circ l$ and $Q \circ n$ for some point Q.

It is noted that, for $l, m \in \mathscr{L}$, exactly one of the following holds:

(a) $l = m$;
(b) $l \parallel m$;
(c) $l \wedge m$ exists and is a point, that is, l and m are distinct nonparallel lines and $l \wedge m$ is the unique point of Theorem 2.1.

This observation and the last theorem give us an intuitive basis for wanting to incorporate the word "plane" in the name adopted for the systems we are studying. There will be no notion of "skew lines."

THEOREM 2.3. If P, Q, R are distinct collinear points, then $P \vee Q = Q \vee R = R \vee P$.

PROOF. Let P, Q, R be distinct collinear points. Then there ex ists a line l such that $P \circ l$, $Q \circ l$, and $R \circ l$. Now, by virtue of the uniqueness in Axiom A_1, $l = P \vee Q, l = Q \vee R$, and $l = R \vee P$. The desired equalities follow.

THEOREM 2.4. If $l, m, n \in \mathscr{L}$ such that $l \parallel m$ and $m \parallel n$, then $l = n$, or $l \parallel n$.

The proof is left as an exercise.

THEOREM 2.5. There exist four points, no three collinear.

PROOF. Let P, Q, R be distinct noncollinear points, guaranteed by Axiom A_3. Then $Q \not\circ P \vee R$ and $R \not\circ P \vee Q$. From Axiom A_2, we have unique lines m and n such that $Q \circ m$, $m \parallel P \vee R$ and $R \circ n$, $n \parallel P \vee Q$. Certainly, $m \neq n$; otherwise, if $m = n$, then $Q \circ n$, contrary to $n \parallel P \vee Q$. In addition, $m \neq P \vee Q$; otherwise, if $m = P \vee Q$, then $P \circ m$ and $P \circ P \vee R$ contrary to $m \parallel P \vee R$. Now, by Theorem 2.2, there exists a point S such that $S \circ m$ and $S \circ n$. By Theorem 2.1, S is unique, that is, $S = m \wedge n$ (see Figure 2.2).

If $S = P$ or $S = R$, then $m \not\parallel P \vee R$, or if $S = Q$, then $n \not\parallel P \vee Q$, both contradictions. Thus S is different from P, Q, R. Now P, R, S are noncollinear; otherwise, if they were collinear, then $S \circ P \vee R$ and $S \circ m$, contrary to $m \parallel P \vee R$. Similarly, we can conclude P, Q, S are noncollinear. Certainly, P, Q, R are noncollinear. Finally, if Q, R, S are collinear, then $m = Q \vee S = R \vee S = n$, contrary to $m \neq n$ Thus Q, R, S are noncollinear. This completes the proof.

Some comments about Figure 2.2 are in order. The figure is not to be considered part of the proof. It was introduced merely to serve as a

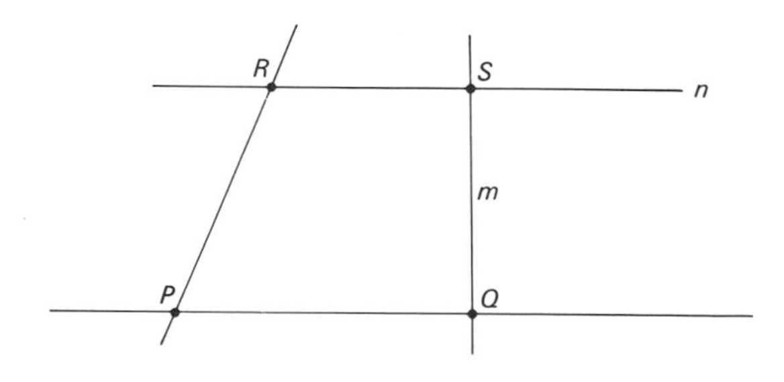

Figure 2.2

catalog of what points and lines were under discussion and what incidence relations they satisfied. Admittedly, a figure might suggest ideas or steps in a proof, but a figure and how it is drawn are not essential to the proof. In fact, how the figure is drawn very well might be misleading. This catalog role of a figure cannot be overemphasized.

THEOREM 2.6. There exist at least two points on every line.

PROOF. Let $l \in \mathscr{L}$. Let P, Q, R, S be distinct points, no three collinear, obtained from Theorem 2.5, where $P \vee Q \parallel R \vee S$ and $P \vee R \parallel Q \vee S$. Now both $l \parallel P \vee Q$ and $l \parallel P \vee R$ cannot hold; otherwise, $P \vee Q = P \vee R$ or $P \vee Q \parallel P \vee R$, both contradictions. Without loss of generality, let $l \nparallel P \vee Q$. In case $l = P \vee Q$, we have $P, Q \circ l$. Now let $l \neq P \vee Q$. Then $P' = l \wedge (P \vee Q)$ exists and $P' \circ l$. We have, also, $l \nparallel R \vee S$. In case $l = R \vee S$, we have $R, S \circ l$. In case $l \neq R \vee S$, $R' = l \wedge (R \vee S)$ exists and $R' \circ l$. Finally, $P' \neq R'$ since $P \vee Q \parallel R \vee S$.

The proofs of the next two theorems are left as exericses.

THEOREM 2.7. For $l, m \in \mathscr{L}$, $l = m$ if and only if $P \circ m$ for every $P \in \mathscr{P}$ such that $P \circ l$.

THEOREM 2.8. Every point is on at least three lines.

EXERCISES

2.1. Verify the following statements that were made without proof in this section:

(a) If l and m are distinct lines such that $l \parallel m$, then $m \parallel l$.

(b) If $l, m \in \mathscr{L}$, then $l = m$ or $l \parallel m$ or $l \wedge m$ exists and is a point, but no two hold.

(c) If P and Q are distinct points, then $P \vee Q = Q \vee P$.

(d) If l and m are distinct nonparallel lines, then $l \wedge m = m \wedge l$.

2.2. Prove Theorem 2.4.

2.3. Let P, Q, R, and S be as in the proof of Theorem 2.5. What can be said about $P \vee S = Q \vee R$, $P \vee S \parallel Q \vee R$ and the existence of $(P \vee S) \wedge (Q \vee R)$ as a point?

2.4. Prove Theorem 2.7.

2.5. Prove Theorem 2.8.

2.6. Let k be an integer with $k \geqq 2$. Show that if some line has exactly k points on it, then every line has exactly k points on it.

2.2. EXAMPLES

We shall consider several examples of affine planes in the remainder of this chapter. This is to help the reader familiarize himself with the axioms and the consequences we have considered. In addition, this is to obtain many different instances of affine planes, thereby illustrating the generality of the theory. Some of the examples will have a technical use later.

The description of each example will begin with the statement of the systems to be used in its construction. The sets $\mathscr{P}$ and $\mathscr{L}$ and the relation $\circ$ are defined in terms of the elements of the given systems. Then the properties of the given systems are used in establishing that the incidence basis $(\mathscr{P}, \mathscr{L}, \circ)$ is an example of an affine plane if, indeed, it is. Not all such incidence bases will necessarily satisfy Axioms A_1–A_3. In fact, as we shall see later, there is a significance ascribed to bases for which this is the case.

An example can be no more meaningful than the common understanding that the reader and author have of the building materials. For this reason we shall restrict our attention to examples constructed from fairly common systems. Of necessity, a considerable bulk of material is assumed in the examples employing euclidean geometry. The reader should not permit himself to become bogged down in these particular examples at the expense of the other matters being considered. These examples are not part of the logical development of the theory and can be skipped without impairing this development.

BASIS 2.1. Let there be given a euclidean plane. The set $\mathscr{P}$ is to consist of the euclidean points and the set $\mathscr{L}$ of the euclidean lines. Incidence is to be taken as the euclidean incidence, that is, for $P \in \mathscr{P}$ and $l \in \mathscr{L}$, $P \circ l$ if P is on l in the euclidean sense.

We see, quickly, that the axioms for an affine plane hold for the basis $(\mathscr{P}, \mathscr{L}, \circ)$. They are, in fact, frequently among the axioms used to characterize a euclidean plane, for example, in Hilbert's "The Foundations of Geometry [12]." Axiom A_2 is usually referred to as "Playfair's form of the parallel postulate," Thus we see that the basis is an affine plane.

BASIS 2.2. Let there be given a tetrahedron, in euclidean space, with vertices A, B, C, and D. The set $\mathscr{P}$ is to consist of the vertices A, B, C, D and the set $\mathscr{L}$ of the six edges AB, AC, AD, BC, BD, CD of the tetrahedron (see Figure 2.3). For $P \in \mathscr{P}$ and $l \in \mathscr{L}$, define $P \circ l$ to mean P is an endpoint of the line segment l. (These terms are taken to be meaningful from our presumed common knowledge of euclidean geometry.)

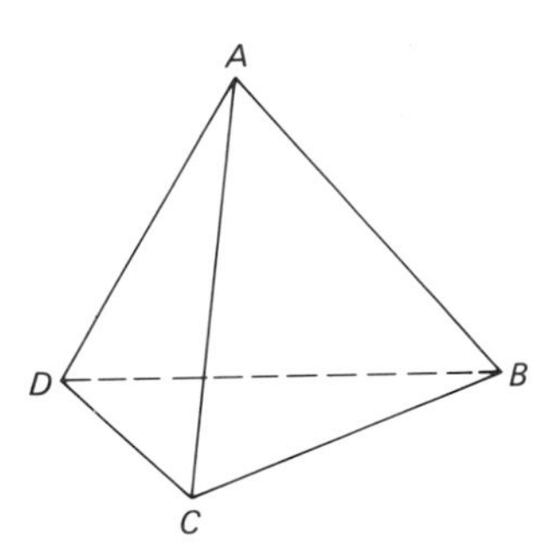

Figure 2.3

The axioms for an affine plane, paraphrased in the euclidean language of this basis, read as follows:

AXIOM A_1. Distinct vertices are the endpoints of one and only one edge.

AXIOM A_2. For a vertex P and edge l such that P is not an endpoint of l, there exists one and only one edge m such that P is an endpoint of m and no vertex is an endpoint of both l and m.

AXIOM A_3. There exist three vertices not all endpoints of one edge.

From our knowledge of tetrahedra in euclidean space, it will follow directly that $(\mathscr{P}, \mathscr{L}, \circ)$ is an affine plane. To illustrate one case when the hypothesis of Axiom A_1 is satisfied, consider distinct points A and B. Then the edge with endpoints A and B would satisfy the requirements demanded of the line in this axiom. To be complete, all such pairs of distinct points would have to be considered. In view of the simplicity of the basis, we think it reasonable to consider Axiom A_1 as having been established.

One case satisfying the hypothesis of Axiom A_2 is $C \not\circ A \vee B$. Then $C \vee D$ is the unique line satisfying $C \circ C \vee D$ and $C \vee D \parallel A \vee B$, that is, the edge with endpoints C and D and the edge with endpoints A and B do not have a vertex as a common endpoint. To be complete, all pairs of points and lines satisfying the hypothesis would have to be considered, but, again, due to the simplicity, we do not deem it necessary to say more concerning Axiom A_2.

For Axiom A_3, we see that A, B, and C are distinct and noncollinear; indeed, no edge of the tetrahedron has more than two endpoints. Thus the basis $(\mathscr{P}, \mathscr{L}, \circ)$ is an affine plane.

BASIS 2.3. Let there be given a tetrahedron, in euclidean space, with vertices W, X, Y, and Z. The set $\mathscr{P}$ is to consist of the four faces WXY, WXZ, WYZ, XYZ and $\mathscr{L}$ of the six edges WX, WY, WZ, XY, XZ, YZ. For $P \in \mathscr{P}$ and $l \in \mathscr{L}$, $P \circ l$, is to mean that P has l as a side.

The discussions of this basis and the next are left as exercises.

BASIS 2.4. Let there be given an octahedron, in euclidean space, with vertices A, B, C, D, E, and F (see Figure 2.4). The set $\mathscr{P}$ is to consist of the four selected faces ABC, ADE, BEF, CDF and $\mathscr{L}$ is to consist of the six vertices. For $P \in \mathscr{P}$ and $l \in \mathscr{L}$, define $P \circ l$ to mean P has l as a vertex.

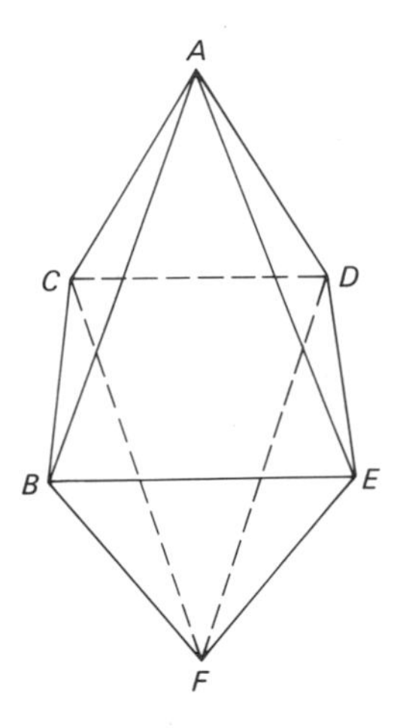

Figure 2.4

For the construction of a "smallest" affine plane, we note that there must exist at least four points P, Q, R, S with no three collinear such that, according to the proof of Theorem 2.5, $P \vee Q \parallel R \vee S$ and $P \vee R \parallel Q \vee S$. Additionally, $P \vee S$ and $Q \vee R$ must exist as lines and, in particular, be distinct and different from the four lines already under consideration. The only way, then, to "generate" additional elements (points or lines) exclusively from these four points and six lines is for the intersection $(P \vee S) \wedge (Q \vee R)$ to exist as a point. We should like, however, to allow for the possibility that $P \vee S \parallel Q \vee R$, thereby limiting this "generation" of additional elements, to provide an abstract basis for the construction of the next example. Hence in constructing the next example, we hypothesize two disjoint sets $\mathscr{P}$ and $\mathscr{L}$ with four and six elements, respectively, and define $\circ$ according to the above possibility.

BASIS 2.5. Let $\mathscr{P} = \{P_1, P_2, P_3, P_4\}$ and $\mathscr{L} = \{l_1, l_2, l_3, l_4, l_5, l_6\}$ such that $P_i \neq P_j$ and $l_i \neq l_j$ for $i \neq j$. Incidence is to be determined from the table in Figure 2.5 as follows: $P_i \circ l_j$ if the symbol "o" is

	l_1	l_2	l_3	l_4	l_5	l_6
P_1	o			o	o	
P_2	o	o				o
P_3		o	o		o	
P^4			o	o		o

Figure 2.5

entered in the row and column corresponding to P_i and l_j, respectively. This device will be called an *incidence table*. Actually, the table also specifies the points and lines.

A basis as simple as the last is sometimes conveniently described, diagrammatically, as in Figure 2.6, where the points are the labeled

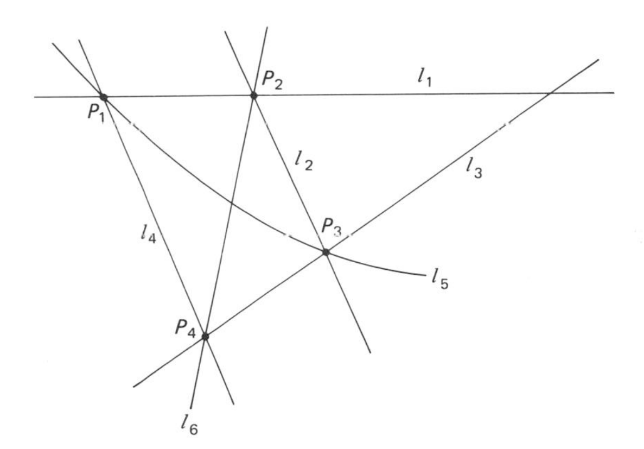

Figure 2.6

"dots," the lines are the labeled "curves," and $P \circ l$ means the dot P is "on" the curve l. This device will be called an *incidence diagram*. Needless to say, this method of specifying the elements of a basis will be only as satisfactory as the clarity with which we are able to label dots and curves, and to draw dots on curves. In the event this clarity is not satisfactory, we shall revert back to the specification of the elements of the basis in the form initially used for Basis 2.5, namely, in the form of an explicit designation of the sets $\mathscr{P}$ and $\mathscr{L}$, and an incidence table. Each of Figures 2.1 and 2.2, referred to in the proofs of Theorems 2.2 and 2.5, can be considered as a portion of an incidence diagram for the affine plane assumed in the hypothesis of the respective proof.

The reader should convince himself that this last basis is an affine plane. A procedure similar to that suggested for Basis 2.2 would have to be followed for a complete analysis.

It can be shown that if an affine plane has more than four points then it has at least nine (see Exercise 2.16). Further, the number of lines would have to be at least 12. The next example has exactly this number of points and lines.

BASIS 2.6. Let $\mathscr{P} = \{P_1, \ldots, P_9\}$ and $\mathscr{L} = \{l_1, \ldots, l_{12}\}$ such that $P_i \neq P_j$ and $l_i \neq l_j$ for $i \neq j$. Incidence is given by the diagram in Figure 2.7.

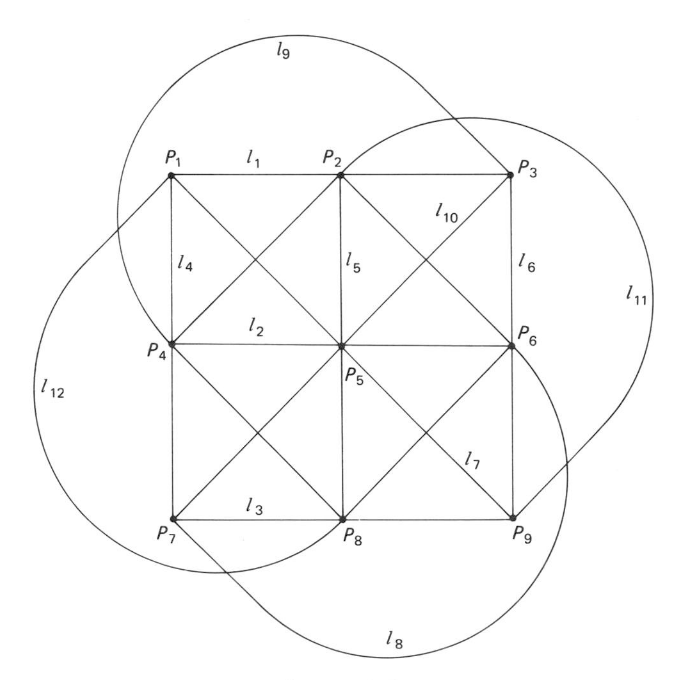

Figure 2.7

A brief discussion of this example is left as an exercise. No attempt will be made to establish fully that it is an affine plane.

The next five examples are algebraic in nature in that the basic systems assumed to be given are algebraic, that is, we begin with number systems.

BASIS 2.7. Consider the real number system. The set $\mathscr{P}$ is to consist of all ordered pairs (x, y) of real numbers and $\mathscr{L}$ of all ordered triples $[a, b, c]$ of real numbers such that not both $a = 0$ and $b = 0$. The triples $[a, b, c]$ and $[a', b', c']$ will be identified as being the same line when $a' = ja$, $b' = jb$, $c' = jc$, for some $j \neq 0$, and this will be indicated by writing $[a, b, c] \equiv [a', b', c']$. Define $(x, y) \circ [a, b, c]$ to mean $ax + by + c = 0$.

In view of the identification of certain triples, it is not clear that incidence is well defined. To this end, let $[a, b, c] \equiv [a', b', c']$. Then $a' = ja$, $b' = jb$, $c' = jc$ for some $j \neq 0$. Now it follows that, for (x, y) a point, $ax + by + c = 0$ if and only if $a'x + b'y + c' = 0$.

Axiom A_1, paraphrased in this example, would read: For (x, y) and (x', y') distinct points, there exist real numbers a, b, c such that

(1) not both $a = 0$ and $b = 0$;
(2) $ax + by + c = 0$ and $ax' + by' + c = 0$;
(3) if a', b', c' are real numbers satisfying
  (a) not both $a' = 0$ and $b' = 0$, and
  (b) $a'x + b'y + c' = 0$ and $a'x' + b'y' + c' = 0$,

then $a' = ja$, $b' = jb$, $c' = jc$, for some $j \neq 0$.

Further discussion of this example is left as an exercise.

BASIS 2.8. This is the same as Basis 2.7 with the real number system replaced by the rational number system.

BASIS 2.9. This is the same as Basis 2.7 with the real number system replaced by the complex number system.

BASIS 2.10. This is the same as Basis 2.7 with the real number system replaced by the number system consisting of {0, 1} with addition and multiplication given by the tables in Figure 2.8.

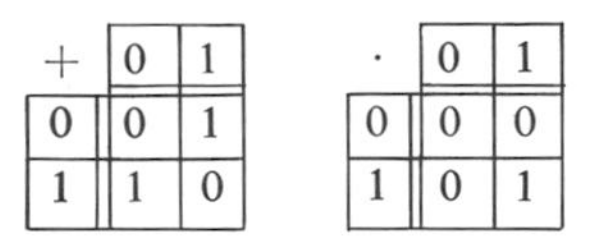

+	0	1
0	0	1
1	1	0

·	0	1
0	0	0
1	0	1

Figure 2.8

BASIS 2.11. This is the same as Basis 2.7 with the real number system replaced by the number system consisting of {0, 1, −1} with addition and multiplication given by the tables in Figure 2.9.

+	0	1	−1
0	0	1	−1
1	1	−1	0
−1	−1	0	1

·	0	1	−1
0	0	0	0
1	0	1	−1
−1	0	−1	1

Figure 2.9

It will be established later that these five bases are affine planes. These five numbers systems have certain common properties of addition and multiplication, and we could, on the basis of these properties alone, show that these algebraic examples are affine planes. In fact, this could be done for any number system with these properties. This is one of the values of considering such systems abstractly. In effect, we shall be doing this later.

The next example will be used for a technical purpose in a later chapter.

BASIS 2.12. Let there be given a rectangular cartesian coordinate system, with axes x and y, in a euclidean plane. Let $\mathscr{P}$ consist of all the euclidean points and $\mathscr{L}$ of the following three types of sets of points:

(1) all the points on the y-axis or on any line parallel to the y-axis;
(2) all the points on any line of nonpositive slope;

(3) all the points with coordinates (u, v) satisfying, for c and d real numbers with $c > 0$,

$$v = \begin{cases} \frac{c}{2}u + d, & u < 0, \\ cu + d, & u \geqq 0. \end{cases}$$

Incidence is to mean "is an element of."

In Figure 2.10, lines l_1, l_2, and l_3 are, respectively, lines of the three types. It is left to an exercise to show that this basis is an affine plane.

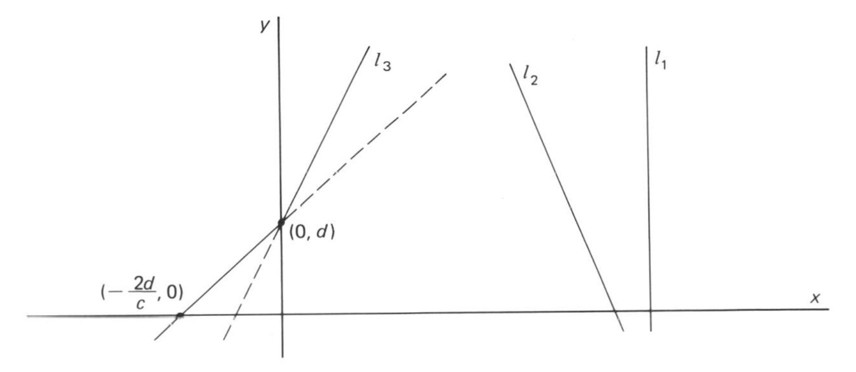

Figure 2.10

A diagrammatic or catalog interpretation has been given to figures drawn on paper. Another interpretation of historic and practical importance is now given. We can interpret

(1) the "dots" made by a pencil (or other marking apparatus) on paper (or other "flat" medium) as points,
(2) the "marks" made by a pencil on paper by running the pencil along a "straightedge" as lines, and
(3) incidence to mean the line "passes through" the point.

Let us suppose we have been given a "sufficiently fine" pencil, a "sufficiently straight" and "long" straightedge, a device for constructing parallel lines (say a parallel ruler used by navigators, a T-square and triangle used by draftsmen, and so on), and a "sufficiently large" paper.

Then we should be inclined to conclude that the points, lines, and incidence described above, insofar as we have experimented, satisfy the axioms of an affine plane within some "degree of accuracy." Essentially, this is what the navigator, draftsman, tinsmith, plumber, surveyor, and so on assume. Actually, they assume much more, namely, that this "geometry" with some additional primitive concepts is a euclidean plane. Indeed, it is reasonable to suppose that through his study of this "geometry" Euclid was led, in part, to his choice of axioms to be assumed and theorems to be proved.

The previous discussion is, certainly, not in keeping with a precise mathematical presentation. Many of the concepts are subject to an individual user's interpretation. Nevertheless, we shall find it instructive to consider this *geometry of the paper* in some of our future discussions. The reader is cautioned not to confuse this use of the paper with the logical development of the mathematical theory. The geometry of the paper will be used only to illustrate and sometimes motivate points of the theory.

We shall have, then, two interpretations for figures drawn on paper: (1) as incidence diagrams for bases (or portions thereof) or (2) as the geometry of the paper, according as the figures are drawn. Whenever the latter interpretation is intended, it will be so stated.

EXERCISES

2.7. In Bases 2.3 and 2.4, state the axioms in the language of the bases. In each find a case when the hypothesis of Axiom A_2 is satisfied and show that the conclusion holds.

2.8. Construct an incidence table for Basis 2.6. What lines are parallel to l_1? To l_8?

2.9. State Axioms A_2 and A_3 in the language of Basis 2.7.

2.10. Construct incidence tables for Bases 2.10 and 2.11.

2.11. Do the "points," "lines," and "on" of euclidean space satisfy the axioms for an affine plane?

2.12. Discuss the axioms in the incidence bases given by the incidence diagrams of Figure 2.11.

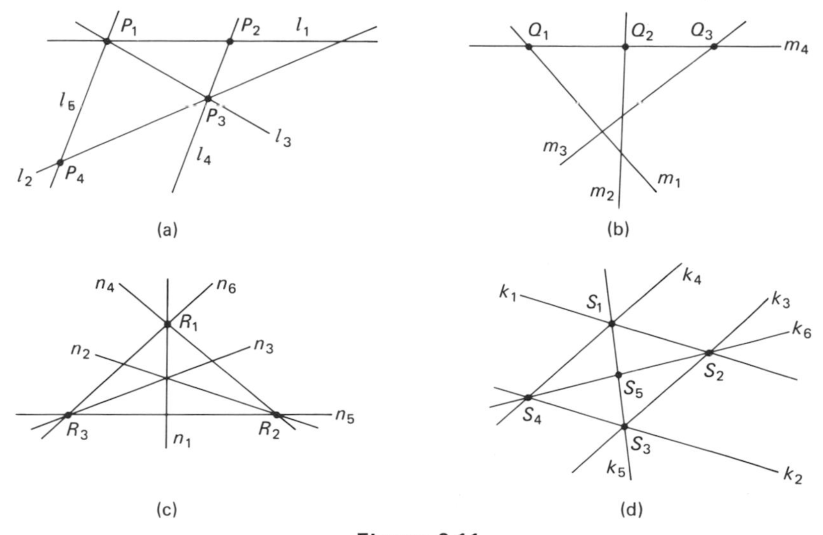

Figure 2.11

2.13. Discuss the axioms in the basis $(\mathscr{P}, \mathscr{L}, \circ)$ where, given a circle in a euclidean plane, we define $\mathscr{P}$ to be the set of euclidean points interior to the circle, $\mathscr{L}$ the set of chords of the circle, and, for $P \in \mathscr{P}$ and $l \in \mathscr{L}$, $P \circ l$ to mean P is on l in the euclidean sense.

2.14. Discuss the axioms in the basis $(\mathscr{P}, \mathscr{L}, \circ)$ where, given a point A in a euclidean plane, we define $\mathscr{P}$ to be the set of euclidean points different from A, $\mathscr{L}$ the set of euclidean lines and circles on A, and, for $P \in \mathscr{P}$ and $l \in \mathscr{L}$, $P \circ l$ to mean P is on l in the euclidean sense.

2.15. Show that Basis 2.12 is an affine plane.

2.16. Prove that, if there exist more than four points in an affine plane, there exist at least nine points.

2.17. Prove that, for an affine plane with exactly three points on some line, the four-point, six-line configuration of Basis 2.5 cannot occur, that is, there cannot be four points A, B, C, D, no three collinear, such that $A \vee B \parallel C \vee D$, $A \vee C \parallel B \vee D$, and $A \vee D \parallel B \vee C$.

PROJECTIVE PLANES

CHAPTER 3

It was noted, in Chapter 1, that parallelism is not invariant under all projections in euclidean space. In view of the objective to state our axioms in terms of concepts that are invariant under all central projections, we shall proceed, at this point, without the notion of parallelism. Thus we shall study incidence bases where, in effect, any two lines are going to intersect. In the next chapter, this new geometry will be related fully to affine planes and later, to euclidean planes.

In this new geometry without parallelism, many statements will be more conveniently stated, where, otherwise, parallel and nonparallel cases would have to be treated separately. Hence many statements will be more concise than they otherwise might be. An additional realization will be the notion of "duality" in which points and lines play symmetric roles. This will lend an additional economy in making statements and constructing proofs.

3.1. AXIOMS FOR A PROJECTIVE PLANE

An incidence basis $(\mathscr{P}, \mathscr{L}, \circ)$ is said to be a *projective plane* if the following are satisfied:

AXIOM P_1. There exist a point P and a line l such that $P \varnothing l$.

AXIOM P_2. Each line has at least three points on it.

AXIOM P_3. If P and Q are distinct points, there exists one and only one line l such that $P \circ l$ and $Q \circ l$.

AXIOM P_4. If l and m are distinct lines, there exists at least one point P such that $P \circ l$ and $P \circ m$.

Throughout this section, it is assumed that $(\mathscr{P}, \mathscr{L}, \circ)$ is a projective plane.

For P and Q distinct points, the unique line l of Axiom P_3 will be called, as in the last chapter, the *join* of P and Q and written $P \vee Q$. As before, we define $P \vee P = P$ and note that $Q \vee P - P \vee Q$.

Before the statement of the first theorems, we should like to extend the meaning of our incidence relation $\circ$ by defining, for $P \in \mathscr{P}$ and $l \in \mathscr{L}$, $l \circ P$ to mean $P \circ l$ and $l \varnothing P$ to mean $P \varnothing l$.

THEOREM 3.1. There exist a line l and a point P such that $l \varnothing P$.

THEOREM 3.2. Each point has at least three lines on it.

THEOREM 3.3. If l and m are distinct lines, there exists one and only one point P such that $l \circ P$ and $m \circ P$.

THEOREM 3.4. If P and Q are distinct points, there exists at least one line l such that $l \circ P$ and $l \circ Q$.

The proofs of the theorems are left as an exercise.

For distinct lines l and m, the unique point P of Theorem 3.3 will be called the *intersection* of l and m and written $l \wedge m$. Again, we define $l \wedge l = l$. Note that $m \wedge l = l \wedge m$.

It is now observed that the statements of the four theorems were obtained from the axioms, respectively, by an interchange of the words "point" and "line" (and an interchange of the corresponding symbols). A statement obtained from another by such an interchange will be called the *dual* of the first; for example, the theorems are, respectively, the dual statements of the axioms, and conversely. For a projective plane, we shall consider the dual statement of a theorem, which has been proved from Axioms P_1–P_4, as having been proved. Indeed, the words "point" and "line" in the proof of the theorem can be interchanged, thereby obtaining a proof of the dual statement based on Theorems 3.1–3.4. This is the *principle of duality* for a projective plane: the dual of any theorem is a theorem.

A set of statements is said to be *self-dual* if it is equivalent to the set of dual statements obtained by dualizing each statement in the given set. This means that the dual of each statement in the given set can be proved using the given set of statements and conversely, each statement in the given set can be proved using the set of dual statements. For example, the set of statements consisting of the four axioms of our present theory is self-dual as is the set consisting of the four theorems. In case a self-dual set consists of exactly one statement, we shall simply say that the statement is *self-dual*. For example, Axiom P_1 is self-dual. (Of course, this is also the case for its dual statement, Theorem 3.1.)

The self-dual nature of a theory, so endowed, is more readily evident when the dual statement of each axiom is in the set of axioms. For example, we could have chosen our axiom set to consist of Axioms P_1–P_4 and Theorems 3.1–3.4. A further refinement is to have each axiom self-dual. This could be realized in the present discussion by letting the axiom set consist of Axiom P_1 plus the conjunctions of Axiom P_2 and Theorem 3.2, of Axiom P_3 and Theorem 3.3, and of Axiom P_4 and Theorem 3.4. These ideas are pursued further in Section 3.4.

"Point" and "line" are considered to be dual concepts and "incidence," as extended in this section, to be self-dual. The definition of each new concept will be accompanied by the definition of its dual concept, unless, of course, it is self-dual. "Join" and "intersection" are dual concepts along with their respective symbols "$\vee$" and "$\wedge$." Two new concepts follow.

Points P, Q, R (lines l, m, n) are said to be *collinear* (*concurrent*) if there exists a line k (a point S) such that $P, Q, R \circ k$ ($l, m, n \circ S$).

In what follows, whenever applicable, the principle of duality will be applied without explicit statement to this effect. In fact, the dual statement to each theorem will not necessarily be stated, but the reader should be aware of such dual statements and their validity.

THEOREM 3.5. If m is a line, there exists a point P such that $P \varnothing m$. Dually, if P is a point, there exists a line m such that $m \varnothing P$.

PROOF. Only the proof of the first part is necessary. Let m be a line. Let P be a point and l a line such that $P \varnothing l$, as guaranteed by Axiom P_1. In case $l = m$, we are finished. Let $l \neq m$. By Axiom P_2, there exists $Q \circ l$ such that $Q \neq l \wedge m$. Now $Q \varnothing m$; otherwise, if $Q \circ m$, we have, by Theorem 3.3, $Q = l \wedge m$ contrary to the above.

THEOREM 3.6. Let P be a point and l a line such that $P \varnothing l$. There exists a one-to-one correspondence between the set of points on l and set of lines on P.

PROOF. Consider P and l with $P \varnothing l$. Define a function π as follows: for $Q \circ l$, $\pi Q = P \vee Q$. That π is the desired one-to-one correspondence is left as an exercise.

COROLLARY 3.7

(a) For l and m lines, there exists a one-to-one correspondence between the set of points on l and the set of points on m.

(b) If there are exactly n points (n a positive integer) on some line, then there are exactly n points on each line.

THEOREM 3.8. Let n be an integer, $n \geqq 2$. If there exist exactly $n + 1$ points on some line, then the projective plane has exactly $n^2 + n + 1$ points.

It should be noted that the theorem does not state that there is a projective plane with exactly $n + 1$ points on some line.

PROOF OF THEOREM 3.8. Let l be a line with exactly $n + 1$ points on it, say $P_1, \ldots, P_{n+1}$, and P a point not on l. Then the lines $P \vee P_1, \ldots, P \vee P_{n+1}$ are distinct, each with $n + 1$ points on it, in particular, each with n points on it different from P. Thus there are at least $(n + 1)n = n^2 + n$ points different from P, that is, at least $n^2 + n + 1$ points in the projective plane. Suppose there is a point Q different from the $n^2 + n + 1$ points just enumerated. Then $Q \varnothing P \vee P_1, \ldots, P \vee P_{n+1}$ since each line has exactly $n + 1$ points on it. Hence $P \vee Q \neq P \vee P_1, \ldots, P \vee P_{n+1}$. From this, we see that P has at least $n + 2$ lines on it. It follows that there cannot be a one-to-one correspondence between the set of points on l and the set of lines on P, contrary to Theorem 3.6. Thus the projective plane has exactly $n^2 + n + 1$ points.

COROLLARY 3.9. There exist at least seven points in a projective plane.

COROLLARY 3.10. If, in a projective plane, there exist more than seven points, then there exist at least 13.

EXERCISES

3.1. Prove Theorems 3.1–3.4.
3.2. Complete the details in the proof of Theorem 3.6.
3.3. Prove Corollary 3.7.
3.4. Prove Corollaries 3.9 and 3.10.
3.5. State the duals to Theorem 3.5–Corollary 3.10. Are any self-dual? Which require proof?
3.6. Let $(\mathscr{P}, \mathscr{L}, \circ)$ be a projective plane. Then $(\mathscr{L}, \mathscr{P}, \circ)$ is an incidence basis. Show that $(\mathscr{L}, \mathscr{P}, \circ)$ is also a projective plane. [It is called the *dual plane* of $(\mathscr{P}, \mathscr{L}, \circ)$.]

3.2. EXAMPLES

BASIS 3.1. Given a point A in euclidean space, let $\mathscr{P}$ consist of the euclidean lines through A and $\mathscr{L}$ of the euclidean planes through A. For $P \in \mathscr{P}$ and $l \in \mathscr{L}$, $P \circ l$ is to mean P is contained in l.

Axiom P_1 reads, paraphrased in the language of the basis, there exist a euclidean line and plane through A such that the line is not contained in the plane. This is certainly the case.

For Axiom P_2, each plane through A contains at least three euclidean lines that pass through A.

Distinct euclidean lines through A are contained in one and only one plane through A. Thus Axiom P_3 is satisfied.

Finally, for Axiom P_4, distinct planes through A contain at least one euclidean line through A, namely, their line of intersection. Thus the incidence basis $(\mathscr{P}, \mathscr{L}, \circ)$ is a projective plane.

BASIS 3.2. Given a sphere in euclidean space, let $\mathscr{P}$ consist of all pairs of antipodal euclidean points on the sphere (that is, endpoints of diameters of the sphere) and $\mathscr{L}$ of the great circles on the sphere. For $P \in \mathscr{P}$ and $l \in \mathscr{L}$, $P \circ l$ is to mean that each euclidean point in the pair P is on the great circle l.

Discussion of this basis is left as an exercise.

We saw, earlier, that any projective plane would have at least three points on each line and at least seven points. We shall construct, next, such a "minimal" example, that is, one with exactly three points on a line and a total of seven points. Of course, dually, there will be exactly three lines on a point and a total of seven lines. A projective plane with exactly seven points and seven lines will be called a *fanian plane* (after Gino Fano, 1871–1952).

BASIS 3.3. Let $\mathscr{P} = \{P_1, \ldots, P_7\}$ and $\mathscr{L} = \{l_1, \ldots, l_7\}$ with $P_i \neq P_j$, $l_i \neq l_j$ for $i \neq j$ and $P_i \neq l_j$. Incidence willl be given by a second form of incidence table. The table in Figure 3.1 is to be read as follows: $P_i \circ l_j$ if P_i appears in the column headed by l_j.

l_1	l_2	l_3	l_4	l_5	l_6	l_7
P_1	P_2	P_3	P_4	P_5	P_6	P_7
P_2	P_3	P_4	P_5	P_6	P_7	P_1
P_4	P_5	P_6	P_7	P_1	P_2	P_3

Figure 3.1

It is immediate that Axioms P_1 and P_2 are satisfied.

For Axiom P_3, first note that each point appears in each row exactly once, in fact, in a definite "cyclic" pattern. Next, note that the three pairs of distinct points in the first column do not occur in any other column. But, by the cyclic nature of the entries in the table, this is true for the three pairs of distinct points in any column, that is, we have the uniqueness part of Axiom P_3. For the existence part, consider P_1. It appears with two additional points in each of three different columns and there are no duplications among these additional points, that is, P_1 appears in some column with each of six additional points (all the points different from P_1). This holds cyclically for each P_i. Hence Axiom P_3 is satisfied.

For Axiom P_4, consider l_1. Each of the points P_1, P_2, P_4 on l_1, appears in two other columns with no two in the same column (other than the first). Hence each of the six columns, different from the first, has one of these points in it, that is, the first column has a point common to each of the other six. But, by the cyclic nature of the table, this is true for each column, and we have Axiom P_4. Thus $(\mathscr{P}, \mathscr{L}, \circ)$ is a projective plane.

Note that, in the table, it was sufficient to have three rows of points each containing every point exactly once such that the uniqueness part of Axiom P_3 was satisfied (the verification of the latter facilitated by the cyclic nature of the table). The first entry in the third row was chosen to be P_4 and the remaining entries determined cyclically so that this property would be realized. Then the rest followed from this property.

From Corollary 3.10, we see that a projective plane with more than seven points would have at least 13. In a projective plane with exactly

13 points, there would be exactly four on each line. The next example is such a projective plane.

BASIS 3.4. Let $\mathscr{P} = \{P_1, \ldots, P_{13}\}$ and $\mathscr{L} = \{l_1, \ldots, l_{13}\}$ with $P_i \neq P_j$ and $l_i \neq l_j$ for $i \neq j$ and $P_i \neq l_j$. An incidence table of the type in the preceding example is to be constructed, but the details are left as an exercise.

We should like, at this time, to raise a question which has been disregarded until now. Consider the situation where, in the study of a theory based on an axiomatic system, a statement and its negation have been proved from the axioms. Both, then, are theorems. Now we are in a position logically to derive any statement as a theorem. This would, indeed, be an uninteresting and unrewarding endeavor. In fact, we should expect that the study would cease and the development be rejected. Such a theory is said to be *inconsistent*. When studying an axiomatic system, how can the question of *consistency* (that is, the freedom from the above anomolous situation) be settled short of obtaining "all possible" theorems and comparing to be sure that none is the negation of any other? The thought of such a procedure is, at best, disheartening. Yet the situation is not as impossible as it might sound. A generally accepted way of determining consistency is to exhibit a basis, called a *consistency basis*, which is an instance of the theory, for if contradictory theorems follow from the axioms, then corresponding contradictory statements must hold in the instance. This says, in effect, that an inconsistent theory cannot have an instance or, stated another way, there are objects like the ones we purport to be studying, a situation greatly to be desired.

There is one catch to this approach. We stressed earlier that a basis can be no more meaningful than the systems used in its construction. This applies equally to the matter of consistency. What we do, in effect, when constructing a consistency basis is to shift the question of consistency of the present theory to the consistency of the systems used in the construction. We have merely passed the buck! We have a *relative* consistency—if the systems used in the construction of the instance are consistent, then the abstract theory is too.

We shall leave the consistency of the given systems an open question or, at least, consider it at a different time.

This does not mean that the question of consistency, necessarily, has an unsatisfactory answer. For example, the materials appealed to in the construction of Basis 3.3 are so primitive, so generally accepted, so "apparently free" of inconsistency, that the resultant incidence basis is highly satisfying considered as an instance of a projective plane. But its freedom from contradiction still depends on the consistency of the notion of "set" and the interpretation of the incidence table in Figure 3.1, something with which few would quarrel, but, nevertheless, something to which the question of consistency has been shifted.

Compare the given materials in Basis 3.3 with those in Bases 3.1 and 3.2, namely, euclidean spaces, the latter being considerably more complex than the theory of projective planes. To be sure, little satisfaction is evoked at this point by knowing that the theory of projective planes is at least as consistent as the theory of euclidean spaces. Obviously, Bases 3.1 and 3.2 were not intended as consistency bases. However, it should be noted that historically one of the ways people began to accept the idea of noneuclidean geometries was exactly through the construction of examples using euclidean geometry.

EXERCISES

3.7. Show that Basis 3.2 is a projective plane. Compare with Basis 3.1.

3.8. The incidence diagram in Figure 3.2 can be used to specify a fanian plane. Label the points and lines in the diagram so that incidence corresponds to that in the table of Basis 3.3.

3.9. Construct an incidence table for Basis 3.4. It should have four rows, each containing every point exactly once such that the uniqueness part of Axiom P_3 is satisfied. Does this imply the rest?

3.10. Let $(\mathscr{P}, \mathscr{L}, \circ)$ be a fanian plane and let $l^* \in \mathscr{L}$. Define $\mathscr{P}' = \{P \in \mathscr{P} : P \not\circ l^*\}$, $\mathscr{L}' = \{l \in \mathscr{L} : l \neq l^*\}$ and, for $P \in \mathscr{P}'$, $l \in \mathscr{L}'$, $P \odot l$ to mean $P \circ l$. What can be said about the incidence basis $(\mathscr{P}', \mathscr{L}', \odot)$? Can your conclusion be generalized?

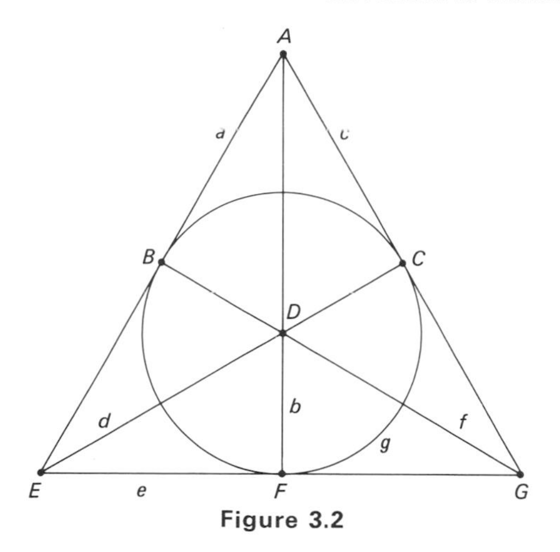

Figure 3.2

3.3. ALGEBRAIC INCIDENCE BASES

BASIS 3.5. Given the real number system, let $\mathscr{P}$ consist of all ordered triples (x, y, z) of real numbers different from $(0, 0, 0)$ and $\mathscr{L}$ of all ordered triples $[a, b, c]$ of real numbers different from $[0, 0, 0]$. (Conceptually, $\mathscr{P}$ and $\mathscr{L}$ are the same. However, for the purpose intended here, they will be considered to be different, in fact, disjoint; and the parenthesis and bracket notations will be employed to help emphasize this.) The triples (x, y, z) and (x', y', z') will be identified as being the same point if $x' = xk, y' = yk, z' = zk$ for some $k \neq 0$, and this will be indicated by writing $(x, y, z) \equiv (x', y', z')$. Dually, $[a, b, c]$ and $[a', b', c']$ are identified if $a' = ja, b' = jb, c' = jc$ for some $j \neq 0$, and this will be written $[a, b, c] \equiv [a', b', c']$. Finally, $(x, y, z) \circ [a, b, c]$ is defined to mean $ax + by + cz = 0$.

To show that incidence is well defined, let $(x, y, z) \equiv (x', y', z')$ and $[a, b, c] \equiv [a', b', c']$. Then $x' = xk, y' = yk, z' = zk$ for some $k \neq 0$, and $a' = ja, b' = jb, c' = jc$ for some $j \neq 0$. Now the following statements are quivalent:

(1) $(x, y, z) \circ [a, b, c]$;

(2) $ax + by + cz = 0$;

(3) $j(ax + by + cz)k = 0$;

(4) $(ja)(xk) + (jb)(yk) + (jc)(zk) = 0$;

(5) $a'x' + b'y' + c'z' = 0$;
(6) $(x', y', z') \circ [a', b', c']$.

Now Basis 3.5 is shown to be a projective plane using only algebraic properties of $+$ and $\cdot$ for the real number system, excluding the "commutativity" of $\cdot$. This is the property that states: for real numbers x and y, $x \cdot y = y \cdot x$. The significance of avoiding the use of the commutativity of multiplication will become evident later. It should be noted that there was a different order of multiplication for the k and j multipliers in the definition of the identification of triples for points and lines.

The proof of Axiom P_1 is left as an exercise. For Axiom P_2, let $[a, b, c] \in \mathscr{L}$. Then not all of a, b, c are 0. It is sufficient to consider the following three cases:

CASE 1. Let $a = b = 0$. Then $(1, 0, 0), (0, 1, 0), (1, 1, 0)$ are distinct and on $[0, 0, c]$.

CASE 2. Let $a = 0$ and $b, c \neq 0$. Then $(1, 0, 0), (0, b^{-1}, -c^{-1})$, $(1, b^{-1}, -c^{-1})$ are distinct and on $[0, b, c]$.

CASE 3. Let $a, b, c \neq 0$. Then $(0, b^{-1}, -c^{-1}), (a^{-1}, 0, -c^{-1})$, $(a^{-1}, -b^{-1}, 0)$ are distinct and on $[a, b, c]$.

For Axiom P_3, let (x_1, y_1, z_1) and (x_2, y_2, z_2) be distinct points. It is necessary to find a line $[a, b, c]$ such that a, b, c satisfy the system of linear equations

$$\begin{aligned} ax_1 + by_1 + cz_1 &= 0, \\ ax_2 + by_2 + cz_2 &= 0 \end{aligned} \tag{1}$$

and the uniqueness property expressed in the identification $\equiv$. Without loss of generality, let $x_1 \neq 0$. Then a, b, c must satisfy

$$a = -by_1x_1^{-1} - cz_1x_1^{-1}.$$

CASE 1. Let $x_2 = 0$. Then b, c must satisfy $by_2 + cz_2 = 0$ and not both $y_2 = 0, z_2 = 0$. Say $y_2 \neq 0$. Then

$$\begin{aligned} b &= -c(z_2y_2^{-1}), \\ a &= c(z_2y_2^{-1}y_1x_1^{-1} - z_1x_1^{-1}) \end{aligned}$$

for any $c \neq 0$ is a solution of (1).

CASE 2. Let $x_2 \neq 0$. The system (1) can be written

$$a(-x_1 x_1^{-1} x_2) + b(-y_1 x_1^{-1} x_2) + c(-z_1 x_1^{-1} x_2) = 0,$$
$$a x_2 + b y_2 + c z_2 = 0.$$

Thus b, c must satisfy $b(y_2 - y_1 x_1^{-1} x_2) + c(z_2 - z_1 x_1^{-1} x_2) = 0$. Now not both $y_2 - y_1 x_1^{-1} x_2 = 0$ and $z_2 - z_1 x_1^{-1} x_2 = 0$; otherwise, if both equal 0, $x_2 = x_1 k, y_2 = y_1 k, z_2 = z_1 k$ with $k = x_1^{-1} x_2 \neq 0$, contrary to the distinctness of (x_1, y_1, z_1) and (x_2, y_2, z_2). Let $y_2 - y_1 x_1^{-1} x_2 \neq 0$. Then $b = -c(z_2 - z_1 x_1^{-1} x_2)(y_2 - y_1 x_1^{-1} x_2)^{-1}$ and $a = c[(z_2 - z_1 x_1^{-1} x_2)(y_2 - y_1 x_1^{-1} x_2)^{-1} y_1 x_1^{-1} - z_1 x_1^{-1}]$ for any $c \neq 0$ is a solution of (1).

For the uniqueness, let $[a', b', c']$ be a line such that

$$a' x_1 + b' y_1 + c' z_1 = 0,$$
$$a' x_2 + b' y_2 + c' z_2 = 0.$$

Then it can be shown, as in the above, if $x_2 = 0$, we have $c' \neq 0$ for $y_2 \neq 0$ and

$$a' = c'(z_2 y_2^{-1} y_1 x_1^{-1} - z_1 x_1^{-1}),$$
$$b' = -c'(z_2 y_2^{-1});$$

and if $x_2 \neq 0$, we have $c' \neq 0$ for $y_2 - y_1 x_1^{-1} x_2 \neq 0$ and

$$a' = c'[(z_2 - z_1 x_1^{-1} x_2)(y_2 - y_1 x_1^{-1} x_2)^{-1} y_1 x_1^{-1} - z_1 x_1^{-1}],$$
$$b' = -c'(z_2 - z_1 x_1^{-1} x_2)(y_2 - y_1 x_1^{-1} x_2)^{-1}.$$

In either case, $j = c' c^{-1}$ is such that $j \neq 0$ and $a' = ja, b' = jb, c' = jc$. Thus $[a', b', c'] \equiv [a, b, c]$.

Verification of Axiom P_4 is left as an exercise.

BASIS 3.6. In Basis 3.5 replace the real number system by the rational number system.

BASIS 3.7. In Basis 3.5 replace the real number system by the complex number system.

BASIS 3.8. In Basis 3.5 replace the real number system by the number system of Basis 2.10.

BASIS 3.9. In Basis 3.5 replace the real number system by the number system of Basis 2.11.

The verification that these four bases are projective planes would read like the verification for Basis 3.5 with the real number system replaced by the appropriate number system. In the verification for Basis 3.5, only properties of addition and multiplication of real numbers were used that hold for the addition and multiplication in the other four systems. These particular properties and number systems with these properties will be identified in Chapter 6 and Appendix B.

We shall refer to Basis 3.5 (or any incidence basis similarly constructed from a given number system) as the *algebraic incidence basis with elements from the real number system* (or *the given number system*).

EXERCISES

3.11. Prove Axioms P_1 and P_4 for Basis 3.5 and complete the details in the justification of Axiom P_2.

3.12. Construct an incidence table for Basis 3.8.

3.13. Do the same for Basis 3.9.

3.14. Let $(\mathscr{P}, \mathscr{L}, \circ)$ be a projective plane. Define, for $l \in \mathscr{L}$, $p(l) = \{P \in \mathscr{P} : P \circ l\}$. Also define $\mathbf{P} = \{\{P\} : P \in \mathscr{P}\}$, $\mathbf{L} = \{p(l) : l \in \mathscr{L}\}$ and, for $Q \in \mathbf{P}$ and $m \in \mathbf{L}$, $Q \mathbf{o}\, m$ to mean $Q \subset m$.

(a) Let $(\mathscr{P}, \mathscr{L}, \circ)$ be Basis 3.3. Construct an incidence table (of the type used for Basis 2.5) for $\mathbf{o}$. Compare with an incidence table for $\circ$.

(b) Show that $(\mathbf{P}, \mathbf{L}, \mathbf{o})$ is a projective plane for $(\mathscr{P}, \mathscr{L}, \circ)$ arbitrary.

3.4. SELF-DUAL AXIOMS

We noted in the first section that the self-duality of a theory is more readily recognizable if the axiom set includes the dual statement of each axiom. This could have been realized for projective planes by letting our axiom set consist of Axioms P_1–P_4 and Theorems 3.1–3.4. We also noted the further refinement where each axiom could be self-dual and we illustrated how this might be realized for our present theory. Both of these choices would be *dependent* sets of axioms, that is, part or all of some axiom can be deduced from the others. It is not incorrect to have a dependent set of axioms; it is merely considered more elegant to have an *independent* set (that is, a set which is not dependent).

Generally, a basis is used to establish that an axiom is not a consequence of the other axioms. The axiom to be shown "independent" must fail in the basis while the others hold. Such an *independence basis* must be constructed for each axiom in order to conclude that the axiom set is independent.

In the next theorem we consider an alternate set of axioms for a projective plane. Each axiom is self-dual and the set is independent. (This set of axioms is, essentially, due to Karl Menger [14], 1948.)

THEOREM 3.11. Let $(\mathscr{P}, \mathscr{L}, \circ)$ be an incidence basis. Then $(\mathscr{P}, \mathscr{L}, \circ)$ is a projective plane if and only if the following are satisfied:

AXIOM I_a. If P and Q are distinct points there exists a line l such that $P \circ l$ and $Q \circ l$.

AXIOM I_b. If l and m are distinct lines there exists a point P such that $l \circ P$ and $m \circ P$.

AXIOM II. There do not exist distinct points P, Q and distinct lines l, m such that $P \circ l$, $Q \circ l$ and $m \circ P$, $m \circ Q$.

AXIOM III. There exist distinct points P, Q and distinct lines l, m such that $P \circ l$, $P \not\circ m$ and $m \circ Q$, $l \not\circ Q$.

AXIOM IV. There exist distinct points P, Q and distinct lines l, m, the points not on the lines, such that some line on both P and Q is on some point on both l and m.

The proof is left as an exercise. The independence of Axioms I_a and I_b can be established using the incidence bases given by the incidence diagrams in Figures 3.3 and 3.4, respectively. In the incidence basis given by Figure 3.3, Axiom I_a fails since $A \neq H$ but there is no line l such that $A, H \circ l$. That the other axioms hold is left as an exercise. In

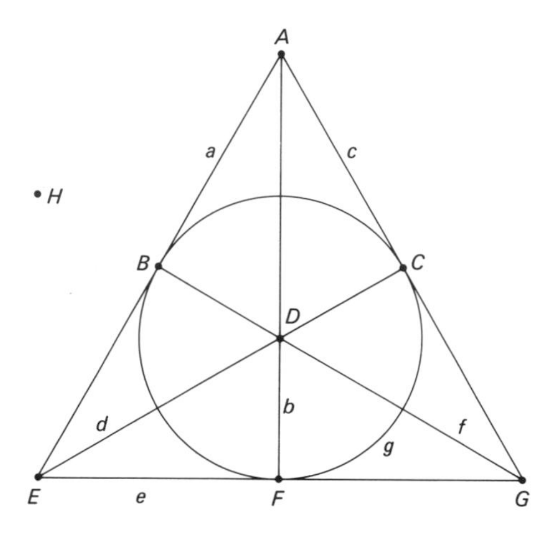

Figure 3.3

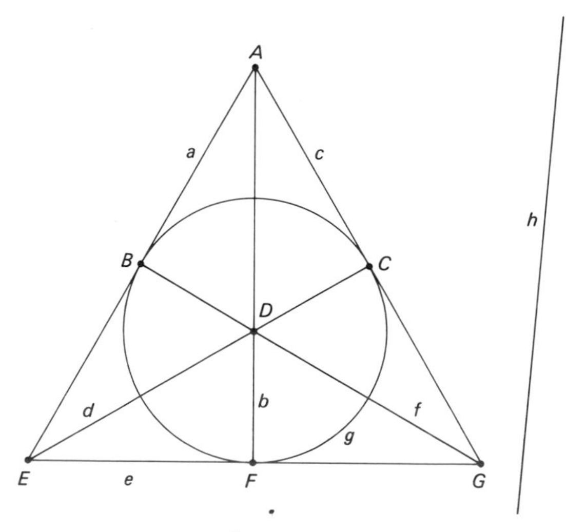

Figure 3.4

addition, the detailed discussion of the remaining independence incidence bases is left as an exercise.

For the independence of Axiom II consider the following: Given a sphere in euclidean space. Let $\mathscr{P}$ consist of the euclidean points on the sphere and $\mathscr{L}$ of the great circles on the sphere. Define incidence to mean euclidean incidence.

Independence incidence bases for Axioms III and IV are given by the incidence diagrams in Figures 3.5 and 3.6, respectively.

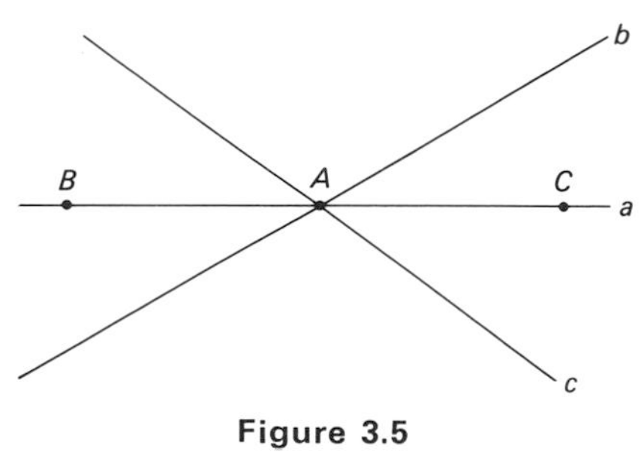

Figure 3.5

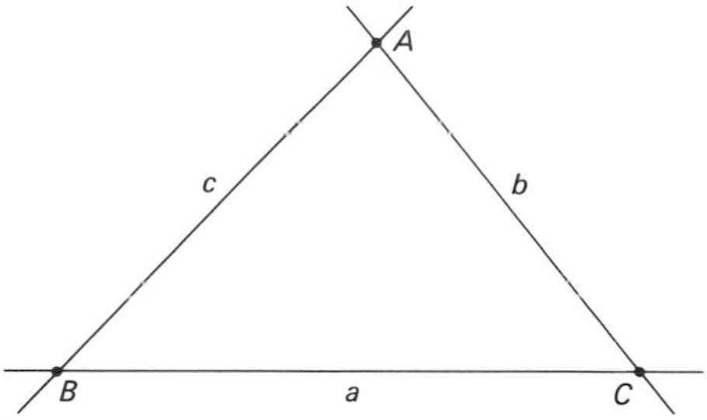

Figure 3.6

EXERCISES

3.15. Prove Theorem 3.11.

3.16. Complete the discussion of the given independence incidence bases.

3.17. Construct a finite independence incidence basis for Axiom II, that is, one in which the sets of points and lines are finite.

3.18 Show that Axioms P_1–P_4 form an independent set, considering Axiom P_3 as two axioms, one for the existence part and one for the uniqueness part. Can this be done with finite independence incidence bases?

AFFINE AND PROJECTIVE PLANES CHAPTER 4

We indicated earlier that the relationship between affine and projective planes would be studied. A hint of what his relationship might be was given in Exercise 3.10, where the deletion of a line and the points on it, in a fanian plane, led to an affine plane. We shall see that such a deletion in any projective plane yields an affine plane. It is natural, then, to inquire whether any affine plane can be obtained from some projective plane as the result of such a deletion. As we shall see there is a sense in which any affine plane can be so realized; moreover, the associated projective plane will be seen to be unique. Thus we shall establish fully the relationship between the theories of affine planes and projective planes, and to study one will be tantamount to studying the other.

The above illustrates an important type of study pursued in mathematics—relating what are apparently different and unrelated theories. The system in one theory is characterized in terms of the system in the other and conversely. Some of the more interesting and important achievements in mathematics have been exactly the discovery of such connecting relationships between theories that had been supposed to be separate and unrelated.

Perhaps analytic geometry is the best-known example of such a discovery. René Descartes (1596–1650), using his invention of coordinate systems, described euclidean geometry algebraically. Through the years his work has undergone considerable refinement and, today, we have a very sophisticated way of viewing his ideas. An indication of this is given in the comment at the end of Section 4.1. In fact, in Chapter 6, we shall undertake a similar study of "analytic geometry" for certain projective planes.

To facilitate our present study we need the concept of isomorphsim, which is the topic of discussion in the next section.

4.1. ISOMORPHISM

The reader, undoubtedly, was aware of some similarity among Bases 2.2–2.5 and 2.10, certainly, of the fact that each consisted of four points and six lines. However, the similarity goes much deeper than merely the number of points and lines. It is possible to construct an incidence table for each in such a way that all the tables would have the same incidence entries, differing only in the labels for the points and lines.

The above is illustrated in Figure 4.1 for Bases 2.2 and 2.3. If the tables

	AB	*BC*	*CD*	*AC*	*BD*	*AD*
A	o			o		o
B	o	o			o	
C		o	o	o		
D			o		o	o

(a) Incidence for Basis 2.2.

	WX	*WZ*	*YZ*	*WY*	*XZ*	*XY*
WXY	o			o		o
WXZ	o	o			o	
WYZ		o	o	o		
XYZ			o		o	o

(b) Incidence for Basis 2.3.

Figure 4.1.

are superimposed the incidence entries are seen to correspond exactly. This suggests that, in some sense, the affine planes differ only in the names given the points and lines. In fact, the superposition suggests how these names might be interchanged (in a one-to-one fashion) for the two bases without changing the incidence.

These ideas are now formulated precisely. This is done, however, in the more general setting of incidence bases, so that these new notions can be applied to affine planes, projective planes, or any geometry with an incidence basis in its foundation.

Let $(\mathscr{P}_1, \mathscr{L}_1, \circ_1)$ and $(\mathscr{P}_2, \mathscr{L}_2, \circ_2)$ be incidence bases. Let π be a one-to-one correspondence between $\mathscr{P}_1$ and $\mathscr{P}_2$ and λ a one-to-one correspondence between $\mathscr{L}_1$ and $\mathscr{L}_2$ such that, for $P \in \mathscr{P}_1$ and $l \in \mathscr{L}_1$, $P \circ_1 l$ if and only if $(\pi P) \circ_2 (\lambda l)$. The ordered pair (π, λ) is said to be an *isomorphism between* $(\mathscr{P}_1, \mathscr{L}_1, \circ_1)$ *and* $(\mathscr{P}_2, \mathscr{L}_2, \circ_2)$, and $(\mathscr{P}_1, \mathscr{L}_1, \circ_1)$ and $(\mathscr{P}_2, \mathscr{L}_2, \circ_2)$ are said to be *isomorphic.*

For example, for Bases 2.5 and 2.4, define π and λ so that $\pi P_1 = ABC$, $\pi P_2 = ADE$, $\pi P_3 = BEF$, $\pi P_4 = CDF$ and $\lambda l_1 = A$, $\lambda l_2 = E$, $\lambda l_3 = F$, $\lambda l_4 = C$, $\lambda l_5 = B$, $\lambda l_6 = D$. Then (π, λ) is an isomorphism between Basis 2.5 and Basis 2.4, and the incidence bases are isomorphic.

We alluded to the notion of isomorphism in Exercise 3.8. In working the exercise the reader went through the essentials of defining an isomorphism between Basis 3.3 and the incidence basis given by the incidence diagram in Figure 3.2.

When incidence bases are affine or projective planes, the conditions of isomorphism can be stated more weakly. This is done in the next two theorems.

THEOREM 4.1. Let $(\mathscr{P}_1, \mathscr{L}_1, \circ_1)$ and $(\mathscr{P}_2, \mathscr{L}_2, \circ_2)$ be affine planes (or projective planes). Let π be a one-to-one correspondence between $\mathscr{P}_1$ and $\mathscr{P}_2$ and λ a one-to-one correspondence between $\mathscr{L}_1$ and $\mathscr{L}_2$ such that, for $P \in \mathscr{P}_1$ and $l \in \mathscr{L}_1$, $P \circ_1 l$ implies $(\pi P) \circ_2 (\lambda l)$. Then (π, λ) is an isomorphism between $(\mathscr{P}_1, \mathscr{L}_1, \circ_1)$ and $(\mathscr{P}_2, \mathscr{L}_2, \circ_2)$.

PROOF. Let $(\mathscr{P}_1, \mathscr{L}_1, \circ_1)$ and $(\mathscr{P}_2, \mathscr{L}_2, \circ_2)$ be affine planes, and π and λ be one-to-one correspondences between $\mathscr{P}_1$ and $\mathscr{P}_2$, and between

$\mathscr{L}_1$ and $\mathscr{L}_2$, respectively, with the stated property. First, we shall prove a statement about joins. Let $\vee_1$ and $\vee_2$ be the joins in $(\mathscr{P}_1, \mathscr{L}_1, \circ_1)$ and $(\mathscr{P}_2, \mathscr{L}_2, \circ_2)$, respectively. If $R, S \in \mathscr{P}_1$ such that $R \neq S$, then $\pi R \neq \pi S$ and πR, $\pi S \circ_2 \lambda(R \vee_1 S)$; thus $\lambda(R \vee_1 S) = (\pi R) \vee_2 (\pi S)$. Now let $P \in \mathscr{P}_1$ and $l \in \mathscr{L}_1$. If $P \circ_1 l$, we have $\pi P \circ_2 \lambda l$. For the converse, let $\pi P \circ_2 \lambda l$. Let $Q \in \mathscr{P}_1$ such that $\pi Q \neq \pi P$ and $\pi Q \circ_2 \lambda l$. (We have at least two points on every line.) Then $\lambda l = (\pi P) \vee_2 (\pi Q)$. Now $P \neq Q$ and $P \circ_1 P \vee_1 Q = \lambda^{-1}(\lambda(P \vee_1 Q)) = \lambda^{-1}((\pi P) \vee_2 (\pi Q)) = \lambda^{-1}(\lambda l) = 1$. Thus (π, λ) is an isomorphism. Note that Axiom A_1 and Theorem 2.6 were the only properties of an affine plane used in the proof. These are properties that hold for a projective plane; hence the proof applies also to the case when the incidence bases are projective planes.

THEOREM 4.2. Let $(\mathscr{P}_1, \mathscr{L}_1, \circ_1)$ and $(\mathscr{P}_2, \mathscr{L}_2, \circ_2)$ be affine planes (or projective planes). Then $(\mathscr{P}_1, \mathscr{L}_1, \circ_1)$ and $(\mathscr{P}_2, \mathscr{L}_2, \circ_2)$ are isomorphic if and only if there exists a one-to-one correspondence π between $\mathscr{P}_1$ and $\mathscr{P}_2$ such that if $P, Q, R \in \mathscr{P}_1$ with P, Q, R collinear, then $\pi P, \pi Q, \pi R$ are collinear.

PROOF. Let $(\mathscr{P}_1, \mathscr{L}_1, \circ_1)$ and $(\mathscr{P}_2, \mathscr{L}_2, \circ_2)$ be affine planes. For the "only if" part, let (π, λ) be an isomorphism between $(\mathscr{P}_1, \mathscr{L}_1, \circ_1)$ and $(\mathscr{P}_2, \mathscr{L}_2, \circ_2)$. Then it is immediate that π will serve as the desired one-to-one correspondence between $\mathscr{P}_1$ and $\mathscr{P}_2$. For the converse, let π be a one-to-one correspondence between $\mathscr{P}_1$ and $\mathscr{P}_2$ with the stated property. Define λ, a function on $\mathscr{L}_1$ to $\mathscr{L}_2$, by the following: For $l \in \mathscr{L}_1$, let P, Q be distinct points on l and define $\lambda l = (\pi P) \vee_2 (\pi Q)$, where $\vee_2$ is the join in $(\mathscr{P}_2, \mathscr{L}_2, \circ_2)$. To show that λ is well-defined (that is, λl defined above is independent of the choice of distinct points P and Q on l), let $l \in \mathscr{L}_1$ and $P, Q, P', Q' \circ_1 l$ such that $P \neq Q$ and $P' \neq Q'$. Then, by hypothesis, $\pi P \neq \pi Q$, $\pi P' \neq \pi Q'$ and $\pi P, \pi Q, \pi P', \pi Q'$ are collinear. Thus $(\pi P) \vee_2 (\pi Q) = (\pi P') \vee_2 (\pi Q')$. Certainly, the domain of λ is $\mathscr{L}_1$ and the range is contained in $\mathscr{L}_2$. To show that the range is all of $\mathscr{L}_2$, let $l_2 \in \mathscr{L}_2$. In addition, let $P_2, Q_2 \circ_2 l_2$ such that $P_2 \neq Q_2$. Now define $P_1 = \pi^{-1} P_2$ and $Q_1 = \pi^{-1} Q_2$. Then $P_1 \neq Q_1$. Define $l_1 = P_1 \vee_1 Q_1$. Then since $P_1, Q_1 \circ_1 l_1$,

$$\lambda l_1 = (\pi P_1) \vee_2 (\pi Q_1) = P_2 \vee_2 Q_2 = l_2 .$$

Hence $\mathscr{L}_2$ is a subset of the range of λ. Thus the range of λ equals $\mathscr{L}_2$. That λ is one-to-one and (π, λ) is the desired isomorphism are left as an exercise.

The next theorem shows the extent to which an isomorphism between two incidence bases reflects the properties of these bases.

THEOREM 4.3. Let $(\mathscr{P}_1, \mathscr{L}_1, \circ_1)$ and $(\mathscr{P}_2, \mathscr{L}_2, \circ_2)$ be isomorphic incidence bases. Then one is an affine (projective) plane if and only if the other is an affine (projective) plane.

PROOF. Let (π, λ) be an isomorphism between incidence bases $(\mathscr{P}_1, \mathscr{L}_1, \circ_1)$ and $(\mathscr{P}_2, \mathscr{L}_2, \circ_2)$. It will be sufficient to consider the "only if" implication for both parts (that is, the affine and projective) of the theorem. For the first, let $(\mathscr{P}_1, \mathscr{L}_1, \circ_1)$ be an affine plane. We shall establish Axiom A_1 for $(\mathscr{P}_2, \mathscr{L}_2, \circ_2)$. To this end, let $P', Q' \in \mathscr{P}_2$ such that $P' \neq Q'$. Then there exist unique $P, Q \in \mathscr{P}_1$ such that $P \neq Q$ and $\pi P = P'$, $\pi Q = Q'$. Let $l = P \vee_1 Q$. Then $P', Q' \circ_2 \lambda l$. To show that λl is unique, let $P', Q' \circ_2 l'$ for $l' \in \mathscr{L}_2$. Then $P, Q \circ_1 \lambda^{-1} l'$. Hence $\lambda^{-1} l' = P \vee_1 Q = l$. Thus $l' = \lambda l$. That Axioms A_2 and A_3 hold for $(\mathscr{P}_2, \mathscr{L}_2, \circ_2)$ and the proof of the projective plane part of the theorem are left as an exercise.

We shall note, at this point, an isomorphism of particular historic significance and, when applied to the geometry of the paper (considered as a euclidean plane), of practical importance. This very deep result of euclidean geometry is the isomorphism of Bases 2.1 and 2.7. (Actually, a stronger form of isomorphism is established, which includes additional primitive notions.) Usually we say we have described the geometry "algebraically" or "analytically." It is essentially this isomorphism that is studied at the beginning of the usual freshman course in plane analytic geometry, but the details are seldom, if ever, exhibited.

EXERCISES

4.1. Find an isomorphism between Bases 2.5 and 2.10. Is there only one?

4.2. Do Exercise 4.1 for Bases 2.6 and 2.11.

4.3. Do Exercise 4.1 for Bases 3.1 and 3.2.

4.4. Do Exercise 4.1 for Bases 3.3 and 3.8.

4.5. Complete the proof of Theorem 4.2.

4.6. In the notation of Exercise 3.14, show that $(\mathscr{P}, \mathscr{L}, \circ)$ is isomorphic to $(\mathbf{P}, \mathbf{L}, \circ)$.

4.7. Complete the proof of Theorem 4.3.

4.8. Show that two affine planes are necessarily isomorphic if both have exactly (a) four points, (b) nine points.

4.9. In the proof of Theorem 4.1, the following statement was established: For distinct points R, S in $\mathscr{P}_1$, $\lambda(R \vee_1 S) = (\pi R) \vee_2 (\pi S)$. What is the analogous statement for distinct lines l, m in $\mathscr{L}_1$? (Consider both the affine and projective cases.) Verify your statements.

4.2. DELETION SUBGEOMETRIES

In this section $(\mathscr{P}, \mathscr{L}, \circ)$ is assumed to be a projective plane with join $\vee$ and intersection $\wedge$. Let $\mathscr{P}' \subset \mathscr{P}$ and $\mathscr{L}' \subset \mathscr{L}$. Define for $P \in \mathscr{P}'$ and $l \in \mathscr{L}'$, $P \odot l$ to mean $P \circ l$. We shall call $(\mathscr{P}', \mathscr{L}', \odot)$ a *subgeometry* of $(\mathscr{P}, \mathscr{L}, \circ)$.

We want to study a particular type of subgeometry. Let $l^* \in \mathscr{L}$. Define $\mathscr{P}' = \{P \in \mathscr{P} : P \not\circ\, l^*\}$ and $\mathscr{L}' = \{l \in \mathscr{L} : l \neq l^*\}$. We shall call $(\mathscr{P}', \mathscr{L}', \odot)$ the *deletion subgeometry associated with* l^* or, simply, a *deletion subgeometry*. The subgeometry of Exercise 3.10, was a deletion subgeometry.

THEOREM 4.4. A deletion subgeometry of a projective plane is an affine plane.

PROOF. Let $l^* \in \mathscr{L}$ and let $(\mathscr{P}', \mathscr{L}', \odot)$ be the associated deletion subgeometry. For Axiom A_1, let $P, Q \in \mathscr{P}'$ with $P \neq Q$. Then $P \vee Q \neq l^*$. Hence $P \vee Q \in \mathscr{L}'$. Now $P \vee Q$ is the required unique line.

For Axiom A_2, let $P \in \mathscr{P}'$, $l \in \mathscr{L}'$ such that $P \not\circ\, l$. [See Figure 4.2, an

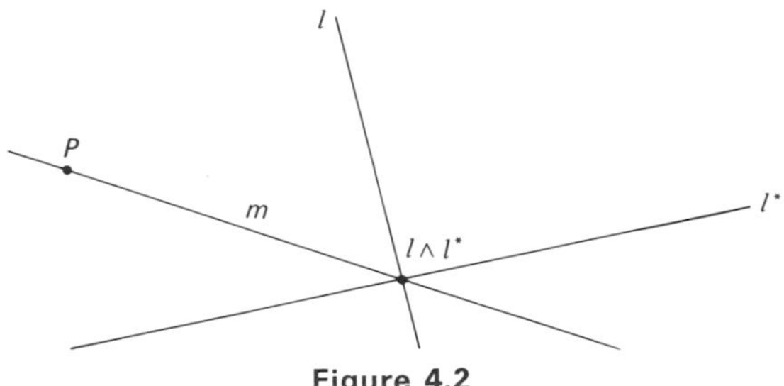

Figure 4.2

incidence diagram in $(\mathscr{P}, \mathscr{L}, \circ)$.] Define $m = P \vee (l \wedge l^*)$. Then $m \in \mathscr{L}'$ and $P \odot m$. In addition, $(m \wedge l) \circ l^*$. Hence $m \parallel l$ in the subgeometry, that is, $Q \odot m$ and $Q \odot l$ for no point $Q \in \mathscr{P}'$. (The only point Q in $\mathscr{P}$ eligible to satisfy $Q \odot m$ and $Q \odot l$ is exactly $m \wedge l$, but $m \wedge l$ has been deleted, that is, $m \wedge l \notin \mathscr{P}'$.) To show that m is unique, let $n \in \mathscr{L}'$ such that $P \odot n$ and $n \parallel l$. Then $(n \wedge l) \circ l^*$; otherwise, if $(n \wedge l) \not\circ l^*$, we should have $n \wedge l \in \mathscr{P}'$ contrary to $n \parallel l$. Now $n \wedge l = l \wedge l^*$ and $n = P \vee (n \wedge l) = P \vee (l \wedge l^*) = m$.

For Axiom A_3, let P, Q, R be a noncollinear triple in $\mathscr{P}$. If none is on l^*, they form a noncollinear triple in $(\mathscr{P}', \mathscr{L}', \odot)$. The remaining cases are left as an exercise and the proof will be considered complete.

We shall exhibit a connection between Bases 2.7 and 3.5. Let $(\mathscr{P}_1, \mathscr{L}_1, \circ_1)$ be the affine plane of Basis 2.7 and $(\mathscr{P}, \mathscr{L}, \circ)$ be the projective plane of Basis 3.5. Recall that

$\mathscr{P}_1 = \{(x, y) : x, y \text{ are real numbers}\}$;
$\mathscr{L}_1 = \{[a, b, c] : a, b, c \text{ are real numbers, not both } a = 0 \text{ and } b = 0\}$;

for $(x, y) \in \mathscr{P}_1$ and $[a, b, c] \in \mathscr{L}_1$; $(x, y) \circ_1 [a, b, c]$ means

$$ax + by + c = 0;$$

and

$$\mathscr{P} = \{(x, y, z) : x, y, z \text{ are real numbers, not all } 0\};$$
$$\mathscr{L} = \{[a, b, c] : a, b, c \text{ are real numbers, not all } 0\};$$

for $(x, y, z) \in \mathscr{P}$ and $[a, b, c] \in \mathscr{L}$, $(x, y, z) \circ [a, b, c]$ means

$$ax + by + cz = 0.$$

Further, elements $(x, y, z), (x', y', z') \in \mathscr{P}$ are identified [written $(x, y, z) \equiv (x', y', z')$] if $x' = xk, y' = yk, z' = zk$ for some $k \neq 0$. Similarly, $[a, b, c], [a', b', c'] \in \mathscr{L}$ (or $[a, b, c], [a', b', c'] \in \mathscr{L}_1$) are identified (written $[a, b, c] \equiv [a', b', c']$) if $a' = ja, b' = jb, c' = jc$ for some $j \neq 0$. Now let $l^* = [0, 0, 1]$ and $(\mathscr{P}', \mathscr{L}', \odot)$ be its associated deletion subgeometry. Then a point (x, y, z) of $\mathscr{P}$ is in $\mathscr{P}'$ if and only if $z \neq 0$, and a line $[a, b, c]$ of $\mathscr{L}$ is in $\mathscr{L}'$ if and only if not both $a = 0$ and $b = 0$. Define π and λ by the following:

$$\pi(x, y, z) = \left(\frac{x}{z}, \frac{y}{z}\right) \qquad \text{for} \quad (x, y, z) \in \mathscr{P}';$$

$$\lambda[a, b, c] = [a, b, c] \qquad \text{for} \quad [a, b, c] \in \mathscr{L}'.$$

To show that π is a well-defined function, let $(x, y, z), (x', y', z') \in \mathscr{P}'$ such that $(x, y, z) \equiv (x', y', z')$, that is, $x' = xk, y' = yk, z' = zk$ for some $k \neq 0$. Then

$$\begin{aligned} \pi(x, y, z) &= \left(\frac{x}{z}, \frac{y}{z}\right) \\ &= \left(\frac{xk}{zk}, \frac{yk}{zk}\right) \\ &= \pi(xk, yk, zk) \\ &= \pi(x', y', z'). \end{aligned}$$

To show that π is a one-to-one correspondence between $\mathscr{P}'$ and $\mathscr{P}_1$, we see immediately that the domain of π is $\mathscr{P}'$ and the range of π is contained in $\mathscr{P}_1$. Let $(x, y) \in \mathscr{P}_1$. Then $(x, y, 1) \in \mathscr{P}'$ and $\pi(x, y, 1) = (x/1, y/1) = (x, y)$, that is, (x, y) is in the range of π. Hence $\mathscr{P}_1$ equals the range of π. Now let $(x, y, z), (x', y', z') \in \mathscr{P}'$ such that $\pi(x, y, z) = \pi(x', y', z')$, that is, $(x/z, y/z) = (x'/z', y'/z')$. Then $z, z' \neq 0$ and $x/z = x'/z', y/z = y'/z'$. Thus for $k = z'/z$, we have $k \neq 0$ and $x' = xk, y' = yk, z' = zk$. Hence $(x, y, z) \equiv (x', y', z')$. This establishes that π is one-to-one.

Similarly, λ can be shown to be a well-defined one-to-one correspondence between $\mathscr{L}'$ and $\mathscr{L}_1$. Finally, (π, λ) is an isomorphism between $(\mathscr{P}', \mathscr{L}', \odot)$ and $(\mathscr{P}_1, \mathscr{L}_1, \circ_1)$. The proofs of these statements are left as an exercise.

We are able to apply the previous discussion to conclude that Basis 2.7 is an affine plane. We know that Basis 3.5 is a projective plane. By Theorem 4.4, $(\mathscr{P}', \mathscr{L}', \odot)$ is an affine plane and, by Theorem 4.3, $(\mathscr{P}_1, \mathscr{L}_1, \circ_1)$ is an affine plane. This conclusion had not been established in Chapter 2.

We saw above that the affine plane of Basis 2.7 was isomorphic to a deletion subgeometry of some projective plane (namely, the deletion subgeometry of Basis 3.5 associated with [0, 0, 1]). This illustrates the sense in which Theorem 4.4 is considered to have a converse, that is, in the sense that any affine plane can be obtained (isomorphically) as the deletion subgeometry of some projective plane. We shall show this in the next section.

First, we indicate what relationship the elements of a deletion subgeometry and their intrinsic properties (those that make the subgeometry an affine plane) bear to the construction of a projective plane isomorphic to $(\mathscr{P}, \mathscr{L}, \circ)$. When passing from a projective plane to a deletion subgeometry, only one line is deleted while at least three (and, in general, many) points are deleted. It would appear, then, that the lines of the deletion subgeometry "retain more information" from the projective plane than the points do. In a sense, this is the case. Thus in the next theorem, we give a description of $(\mathscr{P}, \mathscr{L}, \circ)$ in terms of the elements of $\mathscr{L}$. (This is the dual of what is done in Exercises 3.14 and 4.6.) Then in the ensuing remarks, we see what relationship the lines of the deletion subgeometry bear to this description of $(\mathscr{P}, \mathscr{L}, \circ)$.

THEOREM 4.5. Let $L(P) = \{l \in \mathscr{L} : P \circ l\}$ for $P \in \mathscr{P}$. Further, let $\mathbf{P} = \{L(P) : P \in \mathscr{P}\}$, $\mathbf{L} = \{\{l\} : l \in \mathscr{L}\}$ and for $Q \in \mathbf{P}$ and $m \in \mathbf{L}$, $Q \mathbin{\boldsymbol{\circ}} m$ mean $Q \supset m$. Then $(\mathbf{P}, \mathbf{L}, \boldsymbol{\circ})$ is a projective plane and $(\mathscr{P}, \mathscr{L}, \circ)$ is isomorphic to $(\mathbf{P}, \mathbf{L}, \boldsymbol{\circ})$.

The proof is left as an exercise.

Let $l^* \in \mathscr{L}$ and let $(\mathscr{P}', \mathscr{L}', \odot)$ be its associated deletion subgeometry. Let us see what relationship the elements of the set $L(P)$ of the theorem, for $P \in \mathscr{P}$, bear to the elements of $\mathscr{L}'$ [see Figure 4.3, an incidence diagram in $(\mathscr{P}, \mathscr{L}, \circ)$]. Specifically, for $P \not\circ l^*$, $L(P) \subset \mathscr{L}'$; and for $Q \circ l^*$, $L(Q) \not\subset \mathscr{L}'$, but $L(Q) \subset \mathscr{L}' \cup \{l^*\}$ (or, equivalently, $L(Q)$

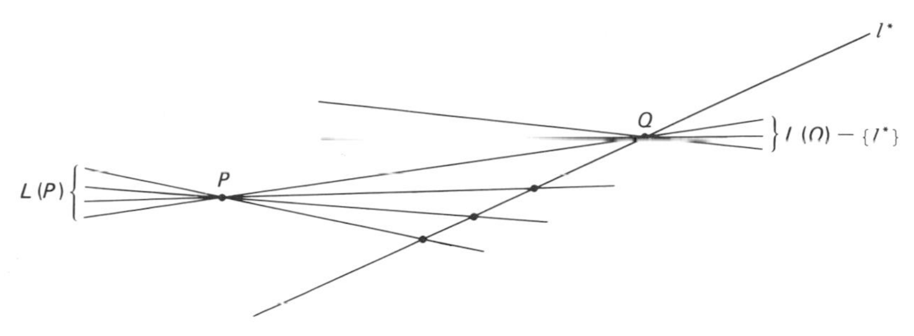

Figure 4.3

$-\{l^*\} \subset \mathscr{L}'$). Thus the points and lines of $(\mathbf{P}, \mathbf{L}, \mathbf{o})$ can be realized as certain subsets of $\mathscr{L}' \cup \{l^*\}$. Further, the lines of the set $L(P)$ satisfy different properties of the incidence $\odot$ in the deletion subgeometry according as $P \not\circ l^*$ or $P \circ l^*$. For $P \not\circ l^*$, the lines of $L(P)$ are concurrent relative to $\odot$. For $Q \circ l^*$, the lines of $L(Q)$ different from l^* (that is, the lines of $L(Q) - \{l^*\}$) are mutually parallel relative to $\odot$. Thus we have focused attention on two classes of subsets of $\mathscr{L}'$, namely, subsets consisting of concurrent lines and subsets consisting of mutually parallel lines. These observations are the basis for the procedure that follows in the next section.

EXERCISES

4.10. Complete the verification of Axiom Λ_3 in the proof of Theorem 4.4.

4.11. Complete the verification of the statements made in the discussion (following the proof of Theorem 4.4) of Bases 3.5 and 2.7. Can this discussion be carried out without using the commutativity of the multiplication of real numbers? How would π and λ have to be defined?

4.12. Consider Theorem 4.5.

(a) Use a fanian plane $(\mathscr{P}, \mathscr{L}, \circ)$ (for example, Basis 3.3) to construct $\mathbf{P}$, $\mathbf{L}$, and an incidence table for $\mathbf{o}$. Establish the conclusion of the theorem.

(b) Prove the theorem in general.

(c) Use $(\mathscr{P}, \mathscr{L}, \circ)$ of part (a) to interpret the comments following the theorem.

4.3. THE IMBEDDING THEOREM

Throughout this section $(\mathscr{P}_1, \mathscr{L}_1, \circ_1)$ is to be an arbitrary affine plane. We shall construct a projective plane $(\mathbf{P}, \mathbf{L}, \circ)$ containing a deletion subgeometry $(\mathbf{P}', \mathbf{L}', \odot)$ isomorphic to the given affine plane $(\mathscr{P}_1, \mathscr{L}_1, \circ_1)$. We shall say that the affine plane is *imbedded*, as a deletion subgeometry, into a projective plane. Further, we shall show that there is essentially one (in the sense of isomorphism) such projective plane.

First, we develop some needed machinery. Join and intersection in $(\mathscr{P}_1, \mathscr{L}_1, \circ_1)$ will be designated by $\vee_1$ and $\wedge_1$. Let l_∞ be an element with the property that $l_\infty \notin \mathscr{L}_1$. (We can choose $l_\infty = \mathscr{L}_1$. Then $l_\infty \notin \mathscr{L}_1$ since, in general, we assume in our use of set theory that $x \in x$ for no element x.) Define, for $l, m \in \mathscr{L}_1$, $l \# m$ to mean $l \parallel m$ or $1 = m$. Then for $m \in \mathscr{L}_1$, define $L(m) = \{l \in \mathscr{L}_1 : l \# m\} \cup \{l_\infty\}$. Also, for $P \in \mathscr{P}_1$, define $L(P) = \{l \in \mathscr{L}_1 : P \circ_1 l\}$.

LEMMA 4.6. Let $P, Q, R \in \mathscr{P}_1$ and $m, n, k \in \mathscr{L}_1$.

(a) $L(P) = L(Q)$ if and only if $P = Q$. Moreover, if $L(P) \neq L(R)$, $P \vee_1 R$ is the only line in $\mathscr{L}_1$ such that $P \vee_1 R \in L(P) \cap L(R)$.

(b) $L(P) \neq L(m)$. Moreover, there exists exactly one line $l \in \mathscr{L}_1$ such that $l \in L(P) \cap L(m)$.

(c) $L(m) = L(n)$ if and only if $m \# n$. Moreover, if $L(m) \neq L(k)$, $j \in L(m)$ and $j \in L(k)$ for no $j \in \mathscr{L}_1$.

PROOF. [See Figure 4.4, an incidence diagram in $(\mathscr{P}_1, \mathscr{L}_1, \circ_1)$.] For part (a), let $P, Q \in \mathscr{P}_1$. Let $L(P) = L(Q)$. Then there exist distinct l, $l' \in \mathscr{L}_1$ such that $P \circ_1 l, l'$. Thus $l, l' \in L(P)$, so $l, l' \in L(Q)$. Hence $Q \circ_1 l, l'$. Now $P = l \wedge l' = Q$. For the converse, $P = Q$ implies $L(P) = L(Q)$ immediately. Finally, for the "moreover" part, let $L(P) \neq L(R)$. Certainly, $P \vee_1 R \in L(P) \cap L(R)$. Let $j \in L(P) \cap L(R)$. Then $j \in L(P), L(R)$. Hence $P, R \circ_1 j$. Thus $j = P \vee_1 R$. The proof of parts (b) and (c) is left as an exercise.

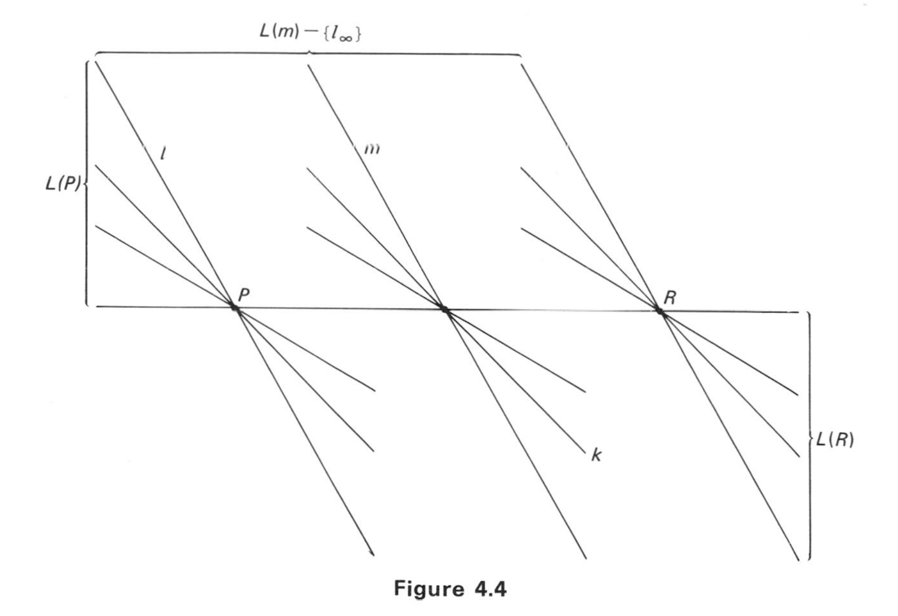

Figure 4.4

THEOREM 4.7. An affine plane can be imbedded as a deletion subgeometry into a projective plane.

PROOF. We shall proceed in the notation introduced prior to the lemma with $(\mathscr{P}_1, \mathscr{L}_1, \circ_1)$ as the given affine plane. Dcfinc

$$\mathbf{P} = \{L(P) : P \in \mathscr{P}_1\} \cup \{L(m) : m \in \mathscr{L}_1\}$$

and

$$\mathbf{L} = \{\{l\} : l \in \mathscr{L}_1 \quad \text{or} \quad l = l_\infty\}.$$

For $Q \in \mathbf{P}$ and $n \in \mathbf{L}$, define $Q \circ n$ to mean $Q \supset n$. First, we prove that $(\mathbf{P}, \mathbf{L}, \circ)$ is a projective plane by establishing Axioms P_1–P_4 [see Exercise 4.14, part (a)].

Let P, Q, $R \in \mathscr{P}_1$ be distinct and noncollinear as guaranteed by Axiom A_3. Then $L(P) \not\supset \{Q \vee_1 R\}$, that is, Axiom P_1 is satisfied.

To show Axiom P_2, let $l' \in \mathbf{L}$. Then $l' = \{l\}$, where $l \in \mathscr{L}_1$ or $l = l_\infty$. In case $l \in \mathscr{L}_1$, let P, Q be distinct points of $\mathscr{P}_1$ on l. Then $L(P)$, $L(Q)$, $L(l)$ are distinct points of $\mathbf{P}$ on l'. In case $l = l_\infty$, let $P \in \mathscr{P}_1$ and m, n,

$k \in \mathscr{L}_1$ be distinct such that $P \circ_1 m, n, k$. Then $L(m), L(n), L(k)$ are distinct points of $\mathbf{P}$ and are on l'. Thus Axiom P_2 holds.

For Axiom P_3, let $P', Q' \in \mathbf{P}$ and be distinct. In case $P' = L(P)$ and $Q' = L(Q)$ for $P, Q \in \mathscr{P}_1, P \neq Q$. Then $\{P \vee_1 Q\}$ is the required line in $\mathbf{L}$, by virtue of part (a) of Lemma 4.6. In case $P' = L(P)$ for $P \in \mathscr{P}_1$ and $Q' = L(m)$ for $m \in \mathscr{L}_1$, apply part (b) of the lemma to obtain the unique $l \in \mathscr{L}_1$ such that $l \in L(P), L(m)$. Then $\{l\}$ is the required line of $\mathbf{L}$. The case $P' = L(m)$ for $m \in \mathscr{L}_1$ and $Q' = L(P)$ for $P \in \mathscr{P}_1$ is similar to the one just considered. For the last case, let $P' = L(m)$ and $Q' = L(n)$ for $m, n \in \mathscr{L}_1$. Then $\{l_\infty\}$ is the required line of $\mathbf{L}$, by virtue of part (c) of the lemma.

Finally, for Axiom P_4, let $m', n' \in \mathbf{L}$ and be distinct. It is sufficient to consider two cases. First, let $m' = \{m\}$ and $n' = \{n\}$ for $m, n \in \mathscr{L}_1$. Then $m \neq n$. If $m \nparallel n$, we have $L(m \wedge_1 n) \circ m', n'$, and if $m \parallel n$, we have $L(m) \circ m', n'$. Secondly, let $m' = \{m\}$ for $m \in \mathscr{L}_1$ and $n' = \{1_\infty\}$. Now $L(m) \circ m', n'$. Thus we can conclude that $(\mathbf{P}, \mathbf{L}, \circ)$ is a projective plane.

Now let $(\mathbf{P}', \mathbf{L}', \odot)$ be the deletion subgeometry of $(\mathbf{P}, \mathbf{L}, \circ)$ associated with $\{1_\infty\}$. Then $\mathbf{P}' = \{L(P) : P \in \mathscr{P}_1\}$ and $\mathbf{L}' = \{\{l\} : l \in \mathscr{L}_1\}$. (Note that for $P' \in \mathbf{P}$, $P' \circ \{1_\infty\}$ if and only if $P' = L(m)$ for some $m \in \mathscr{L}_1$.) That $(\mathscr{P}_1, \mathscr{L}_1, \circ_1)$ is isomorphic to $(\mathbf{P}', \mathbf{L}', \odot)$ is left as an exercise. This completes the proof.

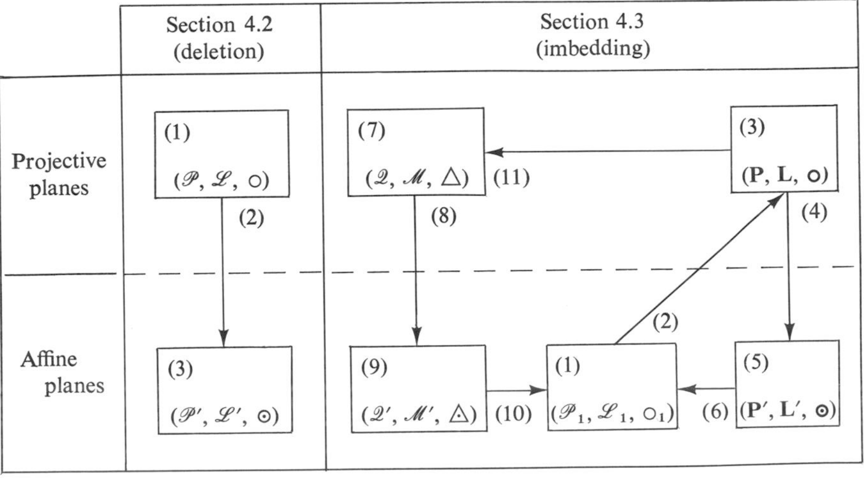

Figure 4.5

Perhaps we should find it helpful to review in block-diagram form the essential results of the last section and those developed to this point in this section (see the table in Figure 4.5). In addition, we shall preview the remaining theorem of this section. The items numbered (1), (2), and (3) under Section 4.2 in the table correspond, respectively, to the given projective plane, the "deletion," and the resultant affine plane of Theorem 4.4. Under Section 4.3 in the table, the items numbered (1)–(6) represent Theorem 4.7, where (1) corresponds to the given affine plane, (2) to the "imbedding," (3) to the resultant projective plane, (4) to the "deletion," (5) to the resultant affine plane, and (6) to the isomorphism between the two affine planes. The items numbered (7)–(11) represent Theorem 4.8 in the following sense: If $(\mathcal{Q}, \mathcal{M}, \triangle)$ is a projective plane with a deletion subgeometry $(\mathcal{Q}', \mathcal{M}', \triangle)$ isomorphic to $(\mathcal{P}_1, \mathcal{L}_1, \circ_1)$, then $(\mathcal{Q}, \mathcal{M}, \triangle)$ is isomorphic to $(\mathbf{P}, \mathbf{L}, \circ)$.

THEOREM 4.8. The projective plane of the last theorem is unique up to isomorphism.

PROOF. Let us continue in the notation of the last theorem. Let $(\mathcal{Q}, \mathcal{M}, \triangle)$ be a projective plane and $(\mathcal{Q}', \mathcal{M}', \triangle)$ the deletion subgeometry associated with $m^* \in \mathcal{M}$. Let (π_0, λ_0) be an isomorphism between $(\mathcal{Q}', \mathcal{M}', \triangle)$ and $(\mathcal{P}_1, \mathcal{L}_1, \circ_1)$. (See the comment immediately preceding the statement of the theorem.) Define, for $P \in \mathcal{Q}$,

$$\pi P = \begin{cases} L(\pi_0 P) & \text{if } P \not\triangle m^*, \\ L(\lambda_0 l) & \text{if } P \triangle m^*, \text{ where } l \in \mathcal{M}' \text{ such that } l \triangle P; \end{cases}$$

and define, for $l \in \mathcal{M}$,

$$\lambda l = \begin{cases} \{\lambda_0 l\} & \text{if } l \neq m^*, \\ \{l_\infty\} & \text{if } l = m^*. \end{cases}$$

(Use the table in Figure 4.5 and thread through items (7)–(10) and (1)–(3) to motivate the definitions of π and λ.) First, we must show that πP is well defined in the case $P \triangle m^*$. To this end, note that there exists $l \in \mathcal{M}$ such that $l \triangle P$ and $l \neq m^*$, that is, $l \in \mathcal{M}'$. Now let $l_1, l_2 \in \mathcal{M}'$ such that $l_1, l_2 \triangle P$ and $l_1 \neq l_2$. Then l_1 and l_2 are parallel relative to $\triangle$, so that $\lambda_0 l_1$ and $\lambda_0 l_2$ are parallel relative to $\circ_1$. Hence $L(\lambda_0 l_1) = L(\lambda_0 l_2)$.

To show that $\mathbf{P}$ is the range of π, let $P' \in \mathbf{P}$. In case $P' = L(P)$ for $P \in \mathscr{P}_1$, $\pi_0^{-1}P \in \mathscr{Q}'$, that is, $\pi_0^{-1}P \not\triangle m^*$. So $\pi(\pi_0^{-1}P) = L(\pi_0(\pi_0^{-1}P)) = L(P) = P'$. In case $P' = L(l)$ for $l \in \mathscr{L}_1$, $\lambda_0^{-1}l \in \mathscr{M}'$. Let $Q \in \mathscr{Q}$ be the intersection of $\lambda_0^{-1}l$ and m^*. Then $\pi Q = L(\lambda_0(\lambda_0^{-1}l)) = L(l) = P'$. Thus the range of π equals $\mathbf{P}$.

To show that π is one-to-one, let $P_1, P_2, \in \mathscr{Q}$ with $\pi P_1 = \pi P_2$. In view of Lemma 4.6 it is sufficient to consider two cases. In case $P_1, P_2, \not\triangle m^*$, $L(\pi_0 P_1) = L(\pi_0 P_2)$, so that $\pi_0 P_1 = \pi_0 P_2$. Thus $P_1 = P_2$. In case $P_1, P_2 \triangle m^*$, let $l_1, l_2 \in \mathscr{M}'$ such that $l_1 \triangle P_1$ and $l_2 \triangle P_2$. Then $L(\lambda_0 l_1) = L(\lambda_0 l_2)$. Thus $\lambda_0 l_1 = \lambda_0 l_2$ or $(\lambda_0 l_1) \parallel (\lambda_0 l_2)$ relative to $\circ_1$; hence $l_1 = l_2$ or l_1 is parallel to l_2 relative to $\triangle$. If $l_1 = l_2$, the intersection of l_1 and m^* is P_1 and P_2, that is, $P_1 = P_2$. If l_1 is parallel to l_2, then l_1, l_2, m^* are concurrent on the point $P_1 = P_2$. Therefore π is a one-to-one correspondence between $\mathscr{Q}$ and $\mathbf{P}$. It is immediate that λ is a one-to-one correspondence between $\mathscr{M}$ and $\mathbf{L}$.

Finally, let $Q \in \mathscr{Q}$ and $m \in \mathscr{M}$ such that $Q \triangle m$. First, consider the case $m \neq m^*$. If $Q \not\triangle m^*$, $Q \in \mathscr{Q}'$ and $(\pi_0 Q) \circ_1 (\lambda_0 m)$, so that $\pi Q = L(\pi_0 Q) \circ \{\lambda_0 m\} = \lambda m$. If $Q \triangle m^*$, let $l \in \mathscr{M}'$ such that $l \triangle Q$. Then $l = m$ or l is parallel to m relative to $\triangle$. From this, we have $\lambda_0 l = \lambda_0 m$ or $(\lambda_0 l) \parallel (\lambda_0 m)$ relative to $\circ_1$. Hence $\pi Q = L(\lambda_0 l) \circ \{\lambda_0 m\} = \lambda m$. For the remaining case, let $m = m^*$. Let $l \in \mathscr{M}'$ such that $l \triangle Q$. Then $\pi Q = L(\lambda_0 l) \circ \{l_\infty\} = \lambda m$. This completes the proof that (π, λ) is an isomorphism between $(\mathscr{Q}, \mathscr{M}, \triangle)$ and $(\mathbf{P}, \mathbf{L}, \circ)$. This completes the proof.

When discussing the projective plane $(\mathbf{P}, \mathbf{L}, \circ)$, as constructed here from the affine plane $(\mathscr{P}_1, \mathscr{L}_1, \circ_1)$, we shall call the sets $L(m)$, with $m \in \mathscr{L}_1$, the *ideal points* (or the *points at infinity*) and the set $\{l_\infty\}$ the *ideal line* (or the *line at infinity*) of $(\mathbf{P}, \mathbf{L}, \circ)$. These ideal elements are exactly the elements of $\mathbf{P}$ and $\mathbf{L}$ not in $\mathbf{P}'$ or $\mathbf{L}'$. It is customary to say that the ideal elements have been appended to $\mathscr{P}_1$ and $\mathscr{L}_1$ and that the incidence $\circ_1$ has been extended to obtain the projective plane $(\mathbf{P}, \mathbf{L}, \circ)$. Certainly, this is not true in a literal sense. However, we do know that $(\mathscr{P}_1, \mathscr{L}_1, \circ_1)$ is isomorphic to $(\mathbf{P}', \mathbf{L}', \odot)$; the ideal elements can be "appended" to $\mathbf{P}'$ and $\mathbf{L}'$ to obtain $\mathbf{P}$ and $\mathbf{L}$; and the incidence $\odot$ can be "extended" to $\circ$, thereby, yielding the projective plane $(\mathbf{P}, \mathbf{L}, \circ)$. This serves to illustrate again the "for all practical purposes" identification role that we let isomorphism play.

EXERCISES

4.13. Complete the proof of Lemma 4.6.

4.14. Consider Theorem 4.7.

(a) Use a four-point, six-line, affine plane $(\mathscr{P}_1, \mathscr{L}_1, \circ_1)$ (for example, Basis 2.5) to construct **P**, **L**, and an incidence table for $\circ$. Show that $(\mathbf{P}, \mathbf{L}, \circ)$ is a projective plane.

(b) Use $(\mathscr{P}_1, \mathscr{L}_1, \circ_1)$ of part (a) to construct $\mathbf{P}'$, $\mathbf{L}'$, and an incidence table for $\odot$. Show that $(\mathscr{P}_1, \mathscr{L}_1, \circ_1)$ is isomorphic to $(\mathbf{P}', \mathbf{L}', \odot)$.

(c) Show that $(\mathscr{P}_1, \mathscr{L}_1, \circ_1)$ is isomorphic to $(\mathbf{P}', \mathbf{L}', \odot)$ in general.

4.15. Show that any two fanian planes are necessarily isomorphic.

4.16. Show that a projective plane with exactly thirteen points cannot contain a subgeometry which is a fanian plane.

4.17. Let n be an integer, $n \geqq 2$. If there exist exactly n points on some line in an affine plane, then the affine plane has exactly n^2 points and $n^2 + n$ lines.

4.18. Let there be given a plane α and a sphere σ in a euclidean space such that α is tangent to σ. Let $(\mathscr{P}_1, \mathscr{L}_1, \circ_1)$ be Basis 2.1 as constructed from α and let $(\mathscr{Q}, \mathscr{M}, \triangle)$ be Basis 3.2 as constructed from σ. Define l^* to be the great circle on σ which lies in a plane parallel to α and $(\mathscr{Q}', \mathscr{M}', \triangle)$ to be the deletion subgeometry of $(\mathscr{Q}, \mathscr{M}, \triangle)$ associated with l^*. Show that $(\mathscr{Q}', \mathscr{M}', \triangle)$ and $(\mathscr{P}_1, \mathscr{L}_1, \circ_1)$ are isomorphic. (*Hint*: Consider "projections" through the center of σ onto α.) Let $(\mathscr{P}_1, \mathscr{L}_1, \circ_1)$ be imbedded into the projective plane $(\mathbf{P}, \mathbf{L}, \circ)$. What elements of $(\mathscr{Q}, \mathscr{M}, \triangle)$ correspond to the ideal elements of $(\mathbf{P}, \mathbf{L}, \circ)$?

4.19. In Figure 4.6, consider the geometry of the paper to be an affine plane $(\mathscr{P}_1, \mathscr{Q}_1, \circ_1)$ imbedded into a projective plane $(\mathbf{P}, \mathbf{L}, \circ)$ (via the construction of this section with $\{l_\infty\}$ as the ideal line). It is given that $m \parallel k^*$ in the figure. Let $(\mathbf{P}', \mathbf{L}', \odot)$ be the deletion subgeometry associated with $\{k^*\}$ and, in this resultant affine plane, designate join, intersection, and parellelism by $\vee$, $\wedge$, and $\Vdash$, respectively. Now draw or indicate the following elements of the set $\mathscr{L}_1 \cup \{l_\infty\}$:

(a) q such that $\{q\} = L(Q) \vee L(n)$;
(b) r' such that $\{r'\} = L(r) \vee L(n)$;
(c) q' such that $L(Q) \odot \{q'\}$ and $\{q'\} \odot \{r\} \wedge \{l_\infty\}$;
(d) r'' such that $L(r'') \odot \{r\}$ and $\{r''\} \odot \{m\} \wedge \{n\}$;
(e) m' such that $L(Q) \odot \{m'\}$ and $\{m'\} \Vert \{m\}$;
(f) n' such that $L(Q) \odot \{n'\}$ and $\{n'\} \Vert \{n\}$;
(g) l' such that $L(Q) \odot (l'\}$ and $\{l'\} \Vert \{l_\infty\}$;
(h) m'' such that $L(r) \odot \{m''\}$ and $\{m''\} \Vert \{m\}$;
(i) n'' such that $L(r) \odot \{n''\}$ and $\{n''\} \Vert \{n\}$.

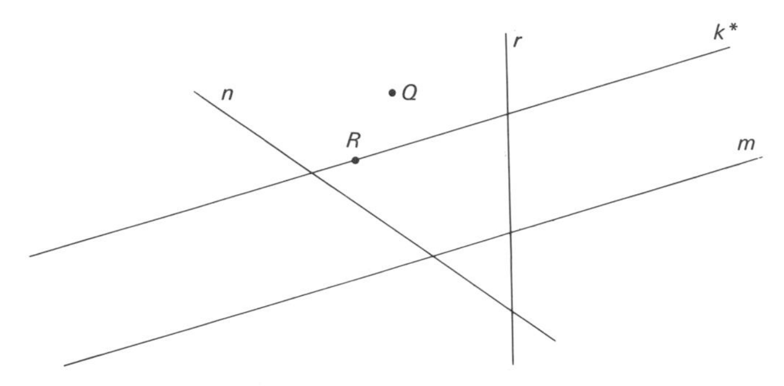

Figure 4.6

4.20. Consider $(\mathscr{P}_1, \mathscr{Q}_1, \circ_1)$ and $(\mathbf{P}, \mathbf{L}, \circ)$ as in the last exercise. However, now define $\mathbf{P}' = \{P \in \mathbf{P} : P \neq L(R)\}$ and $\mathbf{L}' = \{l \in \mathbf{L} : l \varnothing L(R)\}$. Further, for $P \in \mathbf{P}'$ and $l \in \mathbf{L}'$, define $P \odot l$ to mean $P \circ l$. We shall study the geometry associated with the incidence basis $(\mathbf{P}', \mathbf{L}', \odot)$. Define distinct points $A, B \in \mathbf{P}'$ to be *lellarap* if no $c \in \mathbf{L}'$ exists such that $A \odot c$ and $B \odot c$. State three theorems that would characterize this geometry, that is, could be used as axioms for this geometry. Now draw the following elements of $\mathscr{P}_1$ in Figure 4.6:
(a) Q' such that $\{m\} \odot L(Q')$ and $L(Q'), L(Q)$ are lellarap;
(b) N such that $\{m\} \odot L(N)$ and $L(N), L(n)$ are lellarap;
(c) N' such that $\{n\} \odot L\{N'\}$ and $L(N'), L(m)$ are lellarap.
Also draw $k \in \mathscr{L}_1$ such that $\{l_\infty\} \odot L(k)$ and $L(k), L(Q)$ are lellarap. Is k unique?

THEOREMS OF DESARGUES AND PAPPUS

CHAPTER 5

Throughout this chapter we shall assume that $(\mathscr{P}, \mathscr{L}, \circ)$ is a projective plane with join $\vee$ and intersection $\wedge$.

5.1. CONFIGURATIONS

Any nonempty set of points and lines will be called a *configuration*. Certain configurations occur frequently enough to warrant special attention and naming. We shall focus attention on a few in this section.

A set consisting of a triple of noncollinear points (non-concurrent lines) will be called a *triangle* (*trilateral*). Given a triangle (or trilateral) we can consider additional points and lines "generated" by forming joins and intersections of the given elements (or already "generated" elements). In the case of a triangle we would obtain exactly three additional lines. This leads to the next concept. A *complete triangle* (*trilateral*) is a set consisting of the points (lines) of a triangle (trilateral) plus their three distinct joins (intersections). Triangle and trilateral are, of course, dual concepts, as are complete triangle and complete trilateral. However, each of the latter two is self-dual. Hence complete triangle and complete trilateral are the same conceptually. In view of this fact we shall speak only of complete triangles; in fact, we shall refer to them simply as "triangles." Henceforth, when we speak of a triangle, we shall mean a set consisting of three noncollinear points and three nonconcurrent lines such that the lines are pairwise joins of the points (or, equivalently, the points are pairwise intersections of the lines). The points (lines) will be called *vertices* (*sides*). For P, Q, R noncollinear points and l, m, n their pairwise joins, the triangle $\{P, Q, R, l, m, n\}$ will be designated by any of the following symbols: PQR, PRQ, QPR, QRP, RPQ, RQP, lmn, lnm, mln, mnl, nlm, nml.

A set $\{P, Q, R, S\}$, ($\{k, l, m, n\}$) of four distinct points (lines) no three of which are collinear (concurrent) is said to be a *quadrangle* (*quadrilateral*). Let $\{P, Q, R, S\}$ be a quadrangle (see Figure 5.1). Again let us

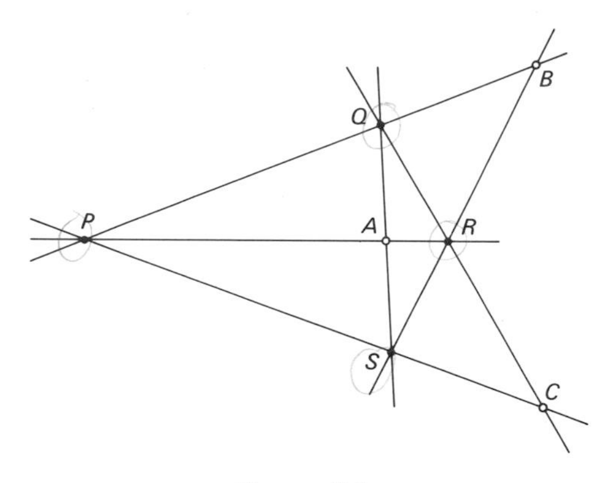

Figure 5.1

consider the additional elements "generated" by these four points. There are the six joins,

$$P \vee Q, \quad P \vee R, \quad P \vee S, \quad Q \vee R, \quad Q \vee S, \quad R \vee S,$$

necessarily distinct, and the three intersections,

$$\begin{aligned} A &= (P \vee R) \wedge (Q \vee S), \\ B &= (P \vee Q) \wedge (R \vee S), \\ C &= (P \vee S) \wedge (Q \vee R). \end{aligned}$$

It can be shown that A, B, C are distinct but it cannot be determined in general whether A, B, C are collinear (see Exercises 5.1 and 5.2). Thus the "generation" of additional elements stops in general. Now the set consisting of P, Q, R, S, the six joins, and A, B, C is said to be a *complete quadrangle*. It will be designated by $PQRS$ or any permutation of the symbols P, Q, R, S written in juxtaposition. The points P, Q, R, S will be called *vertices*, the six lines *sides*, and A, B, C *diagonal points* of the complete quadrangle $PQRS$.

Dually, a quadrilaterial $\{k, l, m, n\}$ gives rise to the *complete quadrilateral* consisting of $k, l, m, n, k \wedge l, k \wedge m, k \wedge n, l \wedge m, l \wedge n, m \wedge n$ and

$$\begin{aligned} x &= (k \wedge l) \vee (m \wedge n), \\ y &= (k \wedge m) \vee (l \wedge n), \\ z &= (k \wedge n) \vee (l \wedge m) \end{aligned}$$

(see Figure 5.2). It will be designated by $klmn$ or any permutation of the

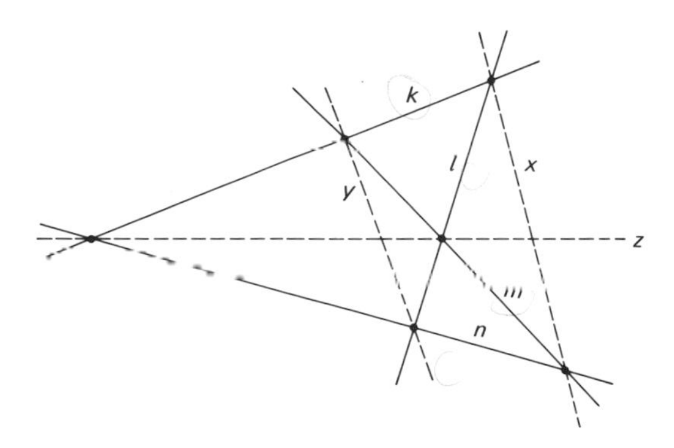

Figure 5.2

symbols k, l, m, n written in juxtaposition. The lines k, l, m, n will be called *sides*, the six points *vertices* and x, y, z *diagonal lines* of the complete quadrilateral $klmn$.

Let PQR and $P'Q'R'$ be triangles with sides l, m, n and l', m', n', respectively. Let $O \in \mathscr{P}$ $(k \in \mathscr{L})$. If $O \circ P \vee P'$, $Q \vee Q'$, $R \vee R'$ $(k \circ l \wedge l', m \wedge m', n \wedge n')$, we shall say that triangles PQR and $P'Q'R'$ are *perspective from the point O (line k)*. The point O (line k)

will be called the *center* (*axis*) of perspectivity. Figure 5.3 illustrates triangles ABC and $A'B'C'$ perspective from O and k.

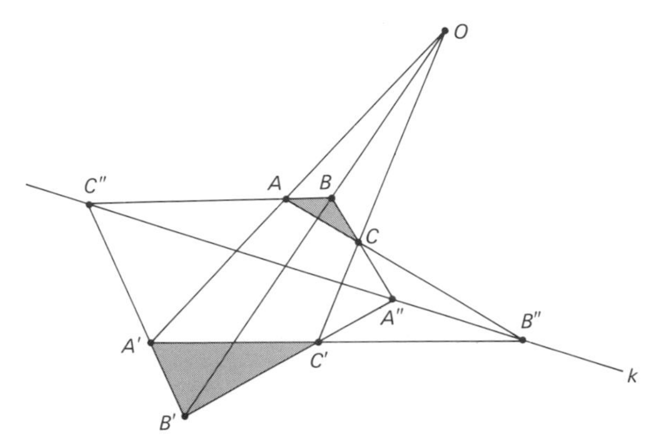

Figure 5.3

EXERCISES

5.1. Show that the diagonal points of a complete quadrangle are necessarily distinct.

5.2. Study the collinearity or noncollinearity of the diagonal points of a complete quadrangle in (a) a fanian plane; (b) a 13-point, 13-line, projective plane; (c) the projective plane obtained by imbedding the geometry of the paper considered as an affine plane (see Exercise 4.19).

5.2. THEOREM OF DESARGUES

THEOREM OF DESARGUES. Triangles perspective from a point are perspective from a line.

This is not a theorem in the sense that it will be proved in a general projective plane. Rather it is an axiom that will be assumed for certain projective planes and the following definition provides us with a convenient device for stating when this assumption is being made. A projective plane is said to be *desarguesian* if the Theorem of Desargues holds and *nondesarguesian* otherwise.

The dual statement of the Theorem of Desargues reads: Triangles perspective from a line are perspective from a point. Note that the dual statement is also the converse statement.

THEOREM 5.1. A desarguesian projective plane is self-dual.

PROOF. It must be shown that in a desarguesian projective plane the dual statement of the Theorem of Desargues holds. Let triangles ABC and $A'B'C'$ be perspective from the line k (see Figure 5.3). Define $A'' = (B \vee C) \wedge (B' \vee C')$, $B'' = (A \vee C) \wedge (A' \vee C')$, $C'' = (A \vee B) \wedge (A' \vee B')$. Then A'', B'', $C'' \circ k$. Now triangles $A'C''A$ and $C'A''C$ are perspective from B''. Hence, by the Theorem of Desargues, they are perspective from a line, that is, $B' = (A' \vee C'') \wedge (C' \vee A'')$, $B = (C'' \vee A) \wedge (A'' \vee C)$, $(A' \vee A) \wedge (C' \vee C)$ are collinear. This states that triangles ABC and $A'B'C'$ are perspective from the point $(A' \vee A) \wedge (C' \vee C)$. This completes the proof.

We should raise the questions of the consistency of a desarguesian projective plane and the independence of the Theorem of Desargues from the other axioms of a desarguesian projective plane. These are answered in the next theorem. Of course, the question of consistency is answered, as before, in a relative way.

THEOREM 5.2. There exist an example of a desarguesian projective plane and an example of a nondesarguesian projective plane.

PROOF. A fanian plane is desarguesian. The details are left as an exercise. A nondesarguesian projective plane is obtained by imbedding Basis 2.12 into a projective plane. The existence of a configuration showing that the Theorem of Desargues fails is left as an exercise. (It should be noted that there are finite nondesarguesian projective planes. More concerning this matter will be stated at the end of Chapter 6.)

The next principal result is to show that Basis 3.5 is desarguesian. The reader is reminded that we shall continue to avoid using the commutativity of the multiplication of real numbers.

LEMMA 5.3. In Basis 3.5, if $P = (x, y, z)$ and $P' = (x', y', z',)$ are distinct points and k, k' are real numbers such that not both $k = 0$ and $k' = 0$, then $P'' = (xk + x'k', yk + y'k', zk + z'k')$ is a point collinear with P and P'.

PROOF. For $k = 0$, $P'' \equiv P'$ and for $k' = 0$, $P'' \equiv P$. Let $k, k' \neq 0$. Suppose P'' is not a point, that is, $P'' = (0, 0, 0)$. Then $xk + x'k' = 0$, $yk + y'k' = 0$ and $zk + z'k' = 0$. Hence $x = x'(-k'k^{-1})$, $y = y'(-k'k^{-1})$, $z = z'(-k'k^{-1})$ contrary to P and P' representing distinct points. Thus P'' is a point. Now let $[a, b, c]$ be the line on P and P'. Then

$$ax + by + cz = 0,$$
$$ax' + by' + cz' = 0.$$

Multiplying the first equation by k, the second by k' and adding, we obtain

$$(ax + by + cz)k + (ax' + by' + cz')k' = 0.$$

Hence $a(xk + x'k') + b(yk + y'k') + c(zk + z'k') = 0$, that is, P'' is on $[a, b, c]$.

LEMMA 5.4. In Basis 3.5, if (x, y, z), (x', y', z'), P are distinct collinear points, there exist real numbers x'', y'', z'', not all zero, and $k'' \neq 0$ such that $P \equiv (x'', y'', z'')$ and $x'' = x + x'k''$, $y'' = y + y'k''$, $z'' = z + z'k''$.

PROOF. Let $P = (x_1, y_1, z_1)$ and let $[a, b, c]$ be the line of collinearity, that is,

$$\begin{aligned} ax + by + cz &= 0, \\ ax' + by' + cz' &= 0, \\ ax_1 + by_1 + cz_1 &= 0. \end{aligned} \tag{1}$$

Without loss of generality let $x \neq 0$. Two cases are considered.

CASE 1. Let $x' = 0$. Then not both $y' = 0$ and $z' = 0$. Let $y' \neq 0$. In addition, let k, k' be real numbers such that

$$xk + x'k' = x_1,$$
$$yk + y'k' = y_1,$$

so that $k = x^{-1}x_1$ and $k' = y'^{-1}(y_1 - yk)$. Now (1) can be written

$$\begin{aligned} a(-xk) + b(-yk) + c(-zk) &= 0, \\ a(-x'k') + b(-y'k') + c(-z'k') &= 0, \\ ax_1 + by_1 + cz_1 &= 0. \end{aligned}$$

We add to obtain

$$a(x_1 - xk - x'k') + b(y_1 - yk - y'k') + c(z_1 - zk - z'k') = 0,$$

that is,

$$c(z_1 - zk - z'k') = 0. \tag{2}$$

Suppose $c = 0$. Then, from (1), $ax' + by' = 0$, that is, $by' = 0$. Hence $b = 0$. Now, from (1), $ax = 0$, so that $a = 0$. This is contrary to $[a, b, c] \neq [0, 0, 0]$. Thus $c \neq 0$. Now $z_1 - zk - z'k' = 0$ follows from (2). We have

$$\begin{aligned} x_1 &= xk + x'k', \\ y_1 &= yk + y'k', \\ z_1 &= zk + z'k'. \end{aligned} \tag{3}$$

Also $k, k' \neq 0$; otherwise, if $k = 0, P \equiv (x', y', z')$, or if $k' = 0$, $P \equiv (x, y, z)$, both contradictions.

CASE 2. Let $x' \neq 0$. Then not both

$$y' - yx^{-1}x' = 0 \quad \text{and} \quad z' - zx^{-1}x' = 0;$$

otherwise, $x' = x(x^{-1}x'), y' = y(x^{-1}x'), z' = z(x^{-1}x')$ with $x^{-1}x' \neq 0$ contrary to hypothesis. Let $y' - yx^{-1}x' \neq 0$. Let k, k' be real numbers such that

$$xk + x'k' = x_1,$$
$$yk + y'k' = y_1,$$

that is,

$$\begin{aligned} -yx^{-1}xk - yx^{-1}x'k' &= -yx^{-1}x_1, \\ yk + y'k' &= y_1. \end{aligned}$$

Then $k' = (y' - yx^{-1}x')^{-1}(y_1 - yx^{-1}x_1)$ and $k = x^{-1}(x_1 - x'k')$. Again we can use (1) to derive (2), that is, $c(z_1 - zk - z'k') = 0$. Suppose $c = 0$. Then, from (1), we have

$$ax + by = 0,$$
$$ax' + by' = 0,$$

that is,

$$-axx^{-1}x' - byx^{-1}x' = 0,$$
$$ax' + by' = 0.$$

Thus $b(y' - yx^{-1}x') = 0$, so that $b = 0$. Now $ax = 0$. Therefore $a = 0$. This contradicts $[a, b, c] \neq [0, 0, 0]$. Hence $c \neq 0$. Now we have (3) as before with $k, k' \neq 0$.

Finally, for both cases, if $x'' = x_1 k^{-1}, y'' = y_1 k^{-1}, z'' = z_1 k^{-1}$, and $k'' = k'k^{-1}$, we have $k'' \neq 0, P \equiv (x'', y'', z'')$ and $x'' = x + x'k''$, $y'' = y + y'k''$, $z'' = z + z'k''$. This completes the proof.

THEOREM 5.5. Basis 3.5 is desarguesian.

PROOF. In Basis 3.5, let triangles ABC and $A'B'C'$ be perspective from Q. Let $Q = (q_1, q_2, q_3)$, $A = (a_1, a_2, a_3)$, $B = (b_1, b_2, b_3)$, $C = (c_1, c_2, c_3)$. By Lemma 5.4, there exist a_i', b_i', c_i' (throughout the proof the index i is to assume the values 1, 2, 3) and nonzero k, m, n such that

$$A' \equiv (a_1', a_2', a_3') \quad \text{with} \quad a_i' = q_i + a_i k,$$
$$B' \equiv (b_1', b_2', b_3') \quad \text{with} \quad b_i' = q_i + b_i m,$$
$$C' \equiv (c_1', c_2', c_3') \quad \text{with} \quad c_i' = q_i + c_i n.$$

Define

$$A'' = (B \vee C) \wedge (B' \vee C'),$$
$$B'' = (A \vee C) \wedge (A' \vee C'),$$
$$C'' = (A \vee B) \wedge (A' \vee B').$$

Now the point given by the triple

$$a_i k + b_i(-m) = (q_i + a_i k) \cdot 1 + (q_i + b_i m)(-1)$$

is on $A \vee B$ and $A' \vee B'$; hence

$$C'' \equiv (a_1 k + b_1(-m), a_2 k + b_2(-m), a_3 k + b_3(-m)).$$

Similarly, B'' is given by the triple $a_i k + c_i(-n)$ and A'' by $c_i n + b_i(-m)$. Now $C'' \circ A'' \vee B''$ since

$$a_i k + b_i(-m) = (c_i n + b_i(-m)) \cdot 1 + (a_i k + c_i(-n)) \cdot 1.$$

Thus ABC and $A'B'C'$ are perspective from $A'' \vee B''$ which shows that Basis 3.5 is desarguesian.

THEOREM 5.6. Let $P_1P_2P_3P_4$ and $Q_1Q_2Q_3Q_4$ be complete quadrangles in a desarguesian projective plane such that $P_i \neq Q_i$ and let $n = 4$, 5, or 6. If exactly n of the inequalities $P_i \vee P_j \neq Q_i \vee Q_j$ for $i < j$ hold and $n - 1$ of the corresponding intersections $(P_i \vee P_j) \wedge (Q_i \vee Q_j)$ are collinear, then the remaining intersection is collinear with them.

PROOF. Let $(\mathscr{P}, \mathscr{L}, \circ)$ be desarguesian and let $P_1P_2P_3P_4$ and $Q_1Q_2Q_3Q_4$ be complete quadrangles with $P_i \neq Q_i$.

CASE 1. Let $n = 6$. Then $P_i \vee P_j \neq Q_i \vee Q_j$ for $i < j$. Without loss of generality, let $(P_i \vee P_j) \wedge (Q_i \vee Q_j) \circ l$ for $i + j < 7$ (see Figure 5.4). Then triangles $P_1P_2P_3$ and $Q_1Q_2Q_3$ are perspective from some point O, and triangles $P_1P_2P_4$ and $Q_1Q_2Q_4$ are perspective from some

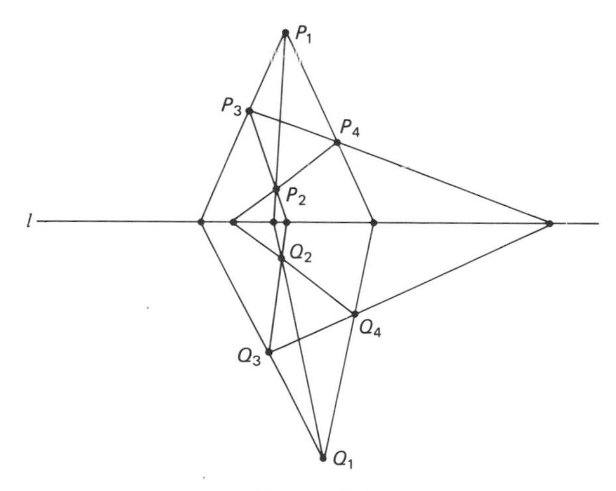

Figure 5.4

point O'. Now $O = (P_1 \vee Q_1) \wedge (P_2 \vee Q_2) = O'$. Hence $P_2 P_3 P_4$ and $Q_2 Q_3 Q_4$ are perspective from the point O, whence from a line, in particular, the line l. Thus $(P_3 \vee P_4) \wedge (Q_3 \vee Q_4) \circ l$. This completes the case for $n = 6$.

CASE 2. Let $n = 5$. Without loss of generality, let $P_1 \vee P_2 = Q_1 \vee Q_2$. There are two essential subcases, namely, when $(P_3 \vee P_4) \wedge (Q_3 \vee Q_4)$ is the "remaining" intersection and when it is not. For the former subcase, let $(P_i \vee P_j) \wedge (Q_i \vee Q_j) \circ l$ for $3 < i + j < 7$ (see Figure 5.5). Define $P = (P_1 \vee P_3) \wedge (P_2 \vee P_4)$ and $Q = (Q_1 \vee Q_3) \wedge$

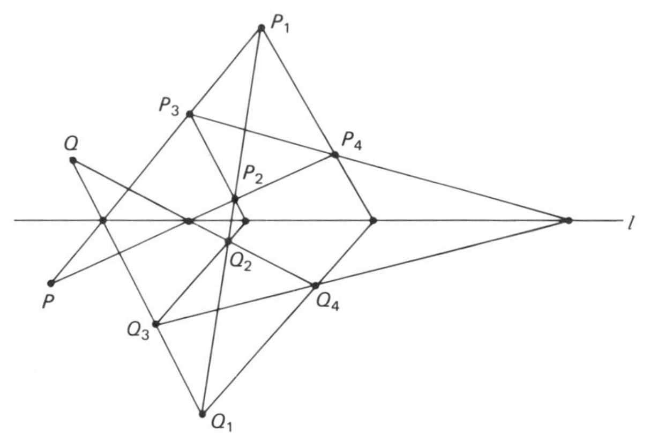

Figure 5.5

$(Q_2 \vee Q_4)$. Then triangles PP_2P_3 and QQ_2Q_3 are perspective from some point O, and triangles PP_1P_4 and QQ_1Q_4 are perspective from some point O'. Now $O = (P \vee Q) \wedge (P_2 \vee Q_2) = (P \vee Q) \wedge (P_1 \vee Q_1) = O'$. Hence $P_1P_3P_4$ and $Q_1Q_3Q_4$ are perspective from the point O, whence from a line, in particular, the line l. So $(P_3 \vee P_4) \wedge (Q_3 \vee Q_4) \circ l$. For the latter subcase, let $(P_i \vee P_j) \wedge (Q_i \vee Q_j) \circ l$ for $i + j > 4$ with no loss of generality. The remainder of this case is left as an exercise.

CASE 3. Let $n = 4$. It is sufficient to consider the case when $P_1 \vee P_2 = Q_1 \vee Q_2$ and $P_3 \vee P_4 = Q_3 \vee Q_4$. The remainder of the proof is left as an exercise.

EXERCISES

5.3. Complete the proof of Theorem 5.2.

5.4. Draw two triangles perspective from two points; from three points.

5.5. Draw a desarguesian configuration with the center of perspectivity on the axis of perspectivity. (This illustrates the *Special Theorem of Desargues* which is given considerable attention in the literature.)

5.6. Let $(\mathbf{P}, \mathbf{L}, \circ)$ be the projective plane obtained by imbedding the geometry of the paper considered as an affine plane $(\mathscr{P}_1, \mathscr{L}_1, \circ_1)$. In this projective plane let triangles ABC and $A'B'C'$ be perspective from the point O and the line l and let $A'' = (B \vee C) \wedge (B' \vee C')$, $B'' = (A \vee C) \wedge (A' \vee C')$, $C'' = (A \vee B) \wedge (A' \vee B')$. Draw a configuration illustrating each of the following cases where ideal elements are specified:

CASE 1. Let $l = \{l_\infty\}$.

(a) $O \not\circ l$; (b) $O \circ l$.

CASE 2. Let $l \neq \{l_\infty\}$.

(a) Only $O \circ \{l_\infty\}$;
(b) only $O, A, A' \circ \{l_\infty\}$;
(c) only $O, A'' \circ \{l_\infty\}$;
(d) $O, A, A', A'' \circ \{l_\infty\}$.

CASE 3. Let $l \neq \{l_\infty\}$ and $O \not\circ \{l_\infty\}$.

(a) No point (of the configuration) on $\{l_\infty\}$;
(b) only $A \circ \{l_\infty\}$;
(c) only $A'' \circ \{l_\infty\}$;
(d) only $A, A'' \circ \{l_\infty\}$;
(e) only $A, B, C'' \circ \{l_\infty\}$;
(f) only $A, B' \circ \{l_\infty\}$.

Cases 2(a) and 3(c) are illustrated in Figures 5.6 and 5.7 where $L(A_1) = A$, $L(B_1) = B$, and so on, and $l = \{l_1\}$.

5.7. Let $(\mathbf{P}, \mathbf{L}, \circ)$ be as in the previous exercise and draw a configuration like that in Figure 5.4 with $l = \{l_\infty\}$. How would this case be stated in affine language?

5.8. State the dual of Theorem 5.6 and draw a configuration for $n = 6$.

5.9. Complete the proof of Theorem 5.6.

5.10. In Theorem 5.6 can the following cases occur and what conclusions would follow?

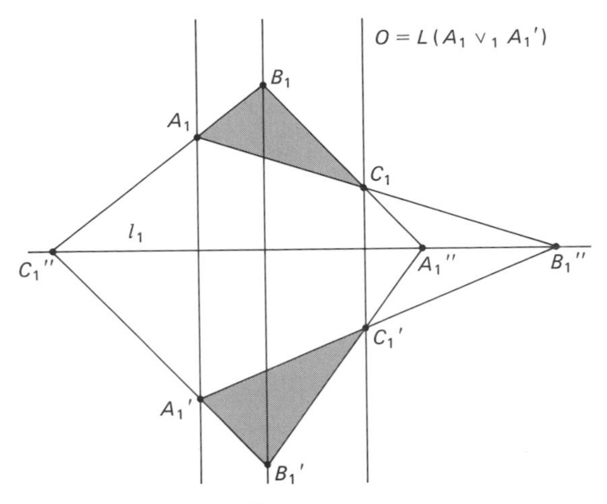

Figure 5.6

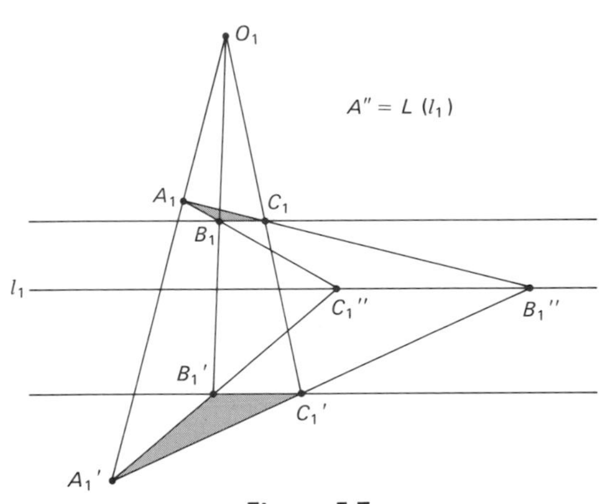

Figure 5.7

(a) $P_1 \vee P_2 = Q_1 \vee Q_2$ and $P_1 \vee P_3 = Q_1 \vee Q_3$.

(b) $P_1 \vee P_2 = Q_1 \vee Q_2$ and $P_1 \vee P_3 = Q_1 \vee Q_4$.

5.11. The only meaningful and significant way to modify $P_i \neq Q_i$ in the hypothesis of Theorem 5.6 is to let $P_i = Q_i$ for exactly one value of i, say $i = 1$, and to let $P_1 \vee P_2 = Q_1 \vee Q_2$, $P_1 \vee P_3 = Q_1 \vee Q_3$, and $P_1 \vee P_4 = Q_1 \vee Q_4$. (Note that n in the theorem necessarily equals 3.) Illustrate this case with an incidence diagram and give it an interpretation.

(When Theorem 5.6 is modified to include this additional possibility, it is sometimes called the *Quadrangle Law* and is, then, equivalent to the Theorem of Desargues.)

5.12. In each part of Figure 5.8 consider the rectangle given to be the outline of your paper. Remaining confined to your paper, use the Theorem of Desargues to draw the indicated line.

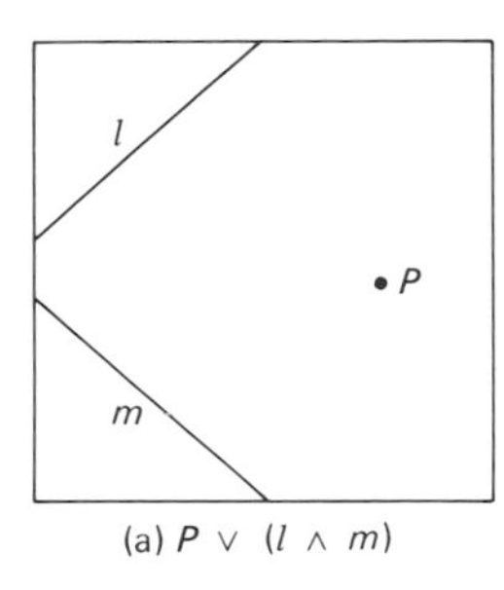

(a) $P \vee (l \wedge m)$

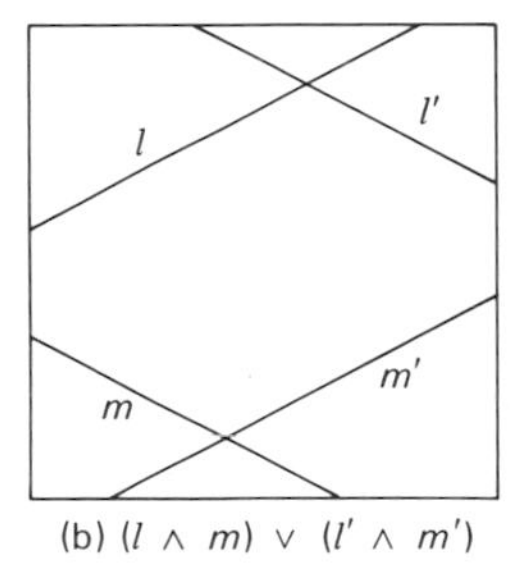

(b) $(l \wedge m) \vee (l' \wedge m')$

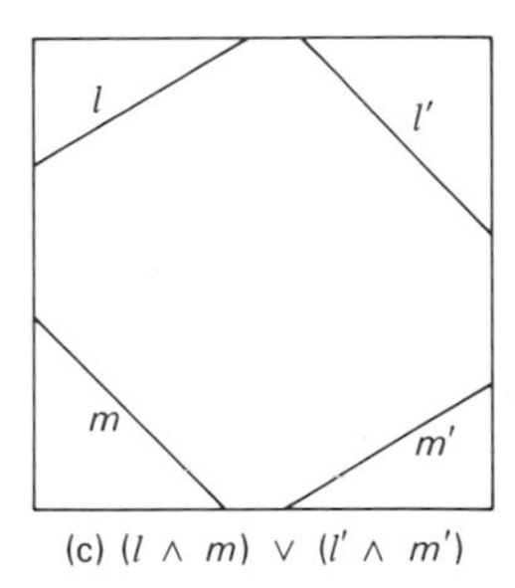

(c) $(l \wedge m) \vee (l' \wedge m')$

Figure 5.8

5.3. THEOREM OF PAPPUS

For an ordered sextuple of points $(P_1, P_2, P_3, P_4, P_5, P_6)$, we shall say that P_i, P_{i+1} for $1 \leqq i \leqq 5$ and P_6, P_1 are *consecutive* points. Now let $(P_1, P_2, P_3, P_4, P_5, P_6)$ be an ordered sextuple of distinct points such that no three consecutive points are collinear. Consider the 12 distinct sextuples obtained by permuting the points cyclically and by reversing the order of the points obtained in each cyclic permutation, that is, the sextuples

$$\begin{array}{ll}
(P_1, P_2, P_3, P_4, P_5, P_6), & (P_6, P_5, P_4, P_3, P_2, P_1), \\
(P_2, P_3, P_4, P_5, P_6, P_1), & (P_1, P_6, P_5, P_4, P_3, P_2), \\
(P_3, P_4, P_5, P_6, P_1, P_2), & (P_2, P_1, P_6, P_5, P_4, P_3), \\
(P_4, P_5, P_6, P_1, P_2, P_3), & (P_3, P_2, P_1, P_6, P_5, P_4), \\
(P_5, P_6, P_1, P_2, P_3, P_4), & (P_4, P_3, P_2, P_1, P_6, P_5), \\
(P_6, P_1, P_2, P_3, P_4, P_5), & (P_5, P_4, P_3, P_2, P_1, P_6).
\end{array}$$

The set consisting of these 12 sextuples will be called a *point-cycle* and designated by $\overline{ABCDEF}$, where (A, B, C, D, E, F) is any one of the 12 sextuples. The points will be called *vertices* and the joins of consecutive pairs of vertices *sides*. The pairs of sides $\{P_1 \vee P_2, P_4 \vee P_5\}$, $\{P_2 \vee P_3, P_5 \vee P_6\}$, and $\{P_3 \vee P_4, P_6 \vee P_1\}$ will be called *opposite*. Finally, the point-cycle $\overline{P_1P_2P_3P_4P_5P_6}$ will be called *pappian* if P_1, P_3, P_5 are collinear and P_2, P_4, P_6 are collinear.

The details of dualizing the above are left as an exercise. We shall consider the notion of *line-cycle* with *side*, *vertex*, *opposite*, and *pappian* as having been defined.

THEOREM OF PAPPUS. In a pappian point-cycle the intersections of opposite sides are collinear.

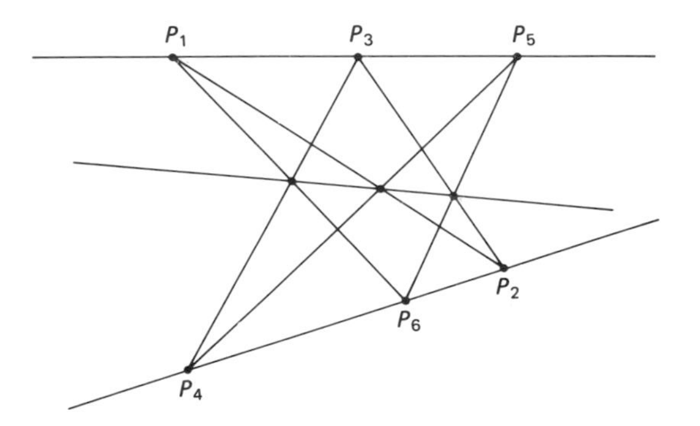

Figure 5.9

As for the Theorem of Desargues, the Theorem of Pappus is an axiom to be assumed for certain projective planes. A projective plane is said to be *pappian* if the Theorem of Pappus holds and *nonpappian* otherwise.

The dual statement of the Theorem of Pappus reads: In a pappian line-cycle the joins of opposite vertices are concurrent.

THEOREM 5.7. A pappian projective plane is self-dual.

The proof is left as an exercise.

THEOREM 5.8. A pappian projective plane is desarguesian.

PROOF. Let $(\mathscr{P}, \mathscr{L}, \circ)$ be pappian. Let triangles PQR and $P'Q'R'$ be perspective from a point O. Define $A = (Q \vee R) \wedge (Q' \vee R')$, $B = (P \vee R) \wedge (P' \vee R')$, $C = (P \vee Q) \wedge (P' \vee Q')$. We shall consider first the case when $P \not\circ Q' \vee R'$ and $Q' \not\circ P \vee R$ (see Figure 5.10). Define

$$\begin{aligned} S &= (P \vee R) \wedge (Q' \vee R'), \\ T &= (P \vee Q') \wedge (R \vee R'), \\ U &= (P \vee Q) \wedge (O \vee S), \\ V &= (P' \vee Q') \wedge (O \vee S). \end{aligned}$$

Then $\overline{PQROSQ'}$ is a pappian point-cycle. Hence

$$\begin{aligned} &(P \vee Q) \wedge (O \vee S) = U, \\ &(Q \vee R) \wedge (S \vee Q') = (Q \vee R) \wedge (Q' \vee R') = A, \\ &(R \vee O) \wedge (Q' \vee P) = (R \vee R') \wedge (P \vee Q') = T \end{aligned}$$

are collinear. Thus $A \circ U \vee T$. In addition, $\overline{Q'P'R'OSP}$ is a pappian point-cycle. Hence

$$\begin{aligned} &(Q' \vee P') \wedge (O \vee S) = V, \\ &(P' \vee R') \wedge (S \vee P) = (P' \vee R') \wedge (P \vee R) = B, \\ &(R' \vee O) \wedge (P \vee Q') = (R \vee R') \wedge (P \vee Q') = T \end{aligned}$$

are collinear. Thus $B \circ T \vee V$.

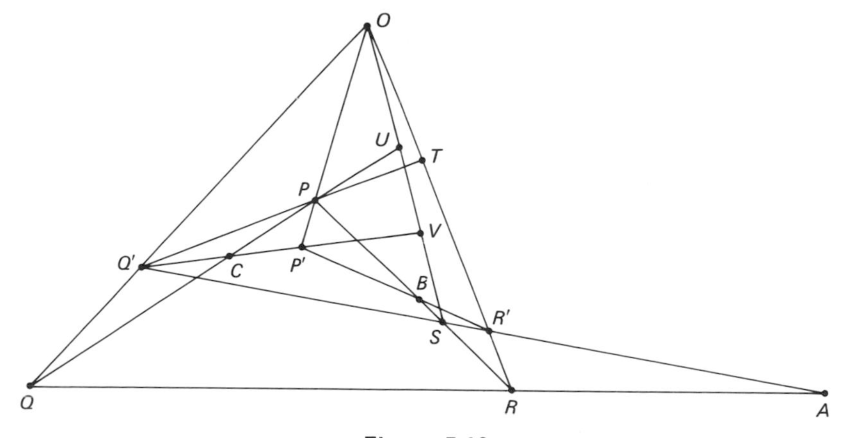

Figure 5.10

Now we shall consider two subcases, namely, when $U \neq V$ and when $U = V$. For the first, we have $\overline{Q'SPUTV}$ a pappian point-cycle. Thus

$$\begin{gathered}(Q' \vee S) \wedge (U \vee T) = (Q' \vee R') \wedge (U \vee T) = A, \\ (S \vee P) \wedge (T \vee V) = (P \vee R) \wedge (T \vee V) = B, \\ (P \vee U) \wedge (V \vee Q') = (P \vee Q) \wedge (P' \vee Q') = C\end{gathered}$$

are collinear. Thus triangles PQR and $P'Q'R'$ are perspective from the line $A \vee B$. For the subcase when $U = V$, we have $P \vee Q$, $P' \vee Q'$, and $O \vee S$ concurrent, so that $U = C = V$ (see Figure 5.11). Now A, B, C,

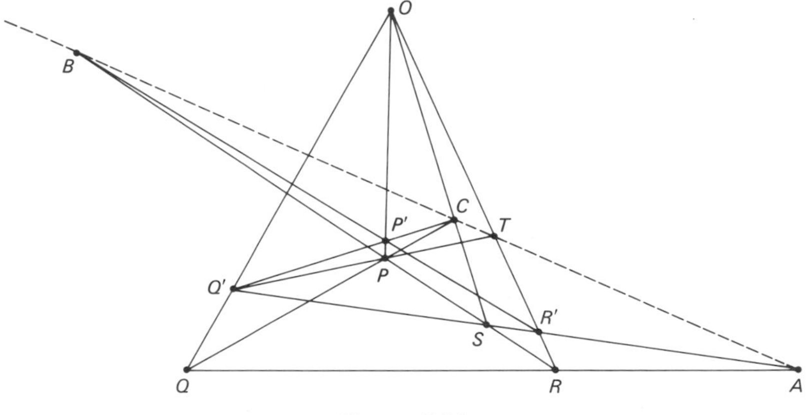

Figure 5.11

T are collinear, that is, triangles PQR and $P'Q'R'$ are perspective from $A \vee B$. The hypothesis $\overline{P \varnothing Q' \vee R'}$ and $\overline{Q' \varnothing P \vee R}$ of the case considered assured us that $\overline{PQROSQ'}$, $\overline{Q'P'R'OSP}$, and $\overline{Q'SPUTV}$ were point-cycles. The remaining case is left as an exercise and we shall consider the proof complete.

THEOREM 5.9. There exist an example of a pappian projective plane and an example of a nonpappian projective plane.

The proof is left as an exercise.

The second part of the theorem can be sharpened considerably; in fact, this is related to the question of whether the converse of Theorem 5.8 is a theorem.

THEOREM 5.10. There exists a desarguesian projective plane which is nonpappian.

PROOF. The construction of such an example is deferred to Chapter 7.

THEOREM 5.11. Basis 3.5 is pappian.

PROOF. In Basis 3.5, let $\overline{AB'CA'BC'}$ be a pappian point-cycle. Define $Q = (A \vee B) \wedge (A' \vee B')$, $X = (A \vee C') \wedge (A' \vee C)$, and $Y = (A \vee B') \wedge (A' \vee B)$. Let $Q = (q_1, q_2, q_3)$, $B = (b_1, b_2, b_3)$, $C' = (c_1', c_2', c_3')$. Then, by Lemma 5.4, there exist a_i, c_i, a_i', b_i' (throughout the proof i is to range over 1, 2, 3) and nonzero j, k, j', k' such that $A \equiv (a_1, a_2, a_3)$, $C \equiv (c_1, c_2, c_3)$, $A' \equiv (a_1', a_2', a_3')$, $B' \equiv (b_1', b_2', b_3')$, and

$$\begin{aligned} a_i &= q_i + b_i j, \\ c_i &= q_i + a_i k, \\ a_i' &= q_i + c_i' k', \\ b_i' &= q_i + a_i' j'. \end{aligned}$$

Thus $b_i' - c_i = a_i'j' - a_i k$. Let $y_i = a_i'j' - b_i j$. Then $y_i = (q_i + a_i'j') - (q_i + b_i j) = b_i' - a_i$, so that $Y \equiv (y_1, y_2, y_3)$. Let $x_i = c_i'k' - a_i k$. Then $x_i = (q_i + c_i'k') - (q_i + a_i k) = a_i' - c_i$. Thus $X \equiv (x_1, x_2, x_3)$. Define $p_i = y_i k + x_i j'$ and $P = (p_1, p_2, p_3)$. Then $P \circ X \vee Y$. Now

$$\begin{aligned} p_i &= (a_i'j' - b_i j)k + (c_i'k' - a_i k)j' \\ &= [(q_i + c_i'k')j' - b_i j]k + [c_i'k' - (q_i + b_i j)k]j' \\ &= c_i'(k'j'k + k'j') - b_i(jk + jkj') + q_i(j'k - kj') \\ &= c_i'(k'j'k + k'j') - b_i(jk + jkj') \end{aligned}$$

since $j'k = kj'$. Thus $P \circ C' \vee B$ since $k'j'k + k'j' \neq 0$ and $jk + jkj' \neq 0$. We have, also,

$$\begin{aligned} p_i &= (a_i'j' - b_i j)k + (c_i'k' - a_i k)j' \\ &= [(b_i' - q_i) - (a_i - q_i)]k + [(a_i' - q_i) - (c_i - q_i)]j' \\ &= (b_i'k - c_i j') + a_i'j' - a_i k \\ &= (b_i'k - c_i j') + (q_i + a_i'j') - (q_i + a_i k) \\ &= b_i'k - c_i j' + b_i' - c_i \\ &= b_i'(k + 1) - c_i(j' + 1). \end{aligned}$$

Thus $P \circ B' \vee C$ since $k + 1 \neq 0$ and $j + 1 \neq 0$. Hence $P = (C' \vee B) \wedge (B' \vee C)$. Since $P \circ X \vee Y$, we see that the intersections of opposite sides are collinear.

It is noted that, in the last proof, the commutativity of multiplication was used when $j'k = kj'$ implied $q_i(j'k - kj') = 0$. This observation is significant for what we do in the next chapter.

EXERCISES

5.13. Define *line-cycle* and related notions.

5.14. Prove Theorem 5.7.

5.15. Complete the proof of Theorem 5.8.

5.16. Prove Theorem 5.9.

5.17. In the proof of Theorem 5.11, why do $k'j'k + k'j' \neq 0$ and $jk + jkj' \neq 0$ hold?

5.18. Show that if, in a pappian projective plane, two triangles are perspective from two points, then they are perspective from a third (see Figure 5.12).

5.19. Show that a 13-point, 13-line, projective plane is pappian.

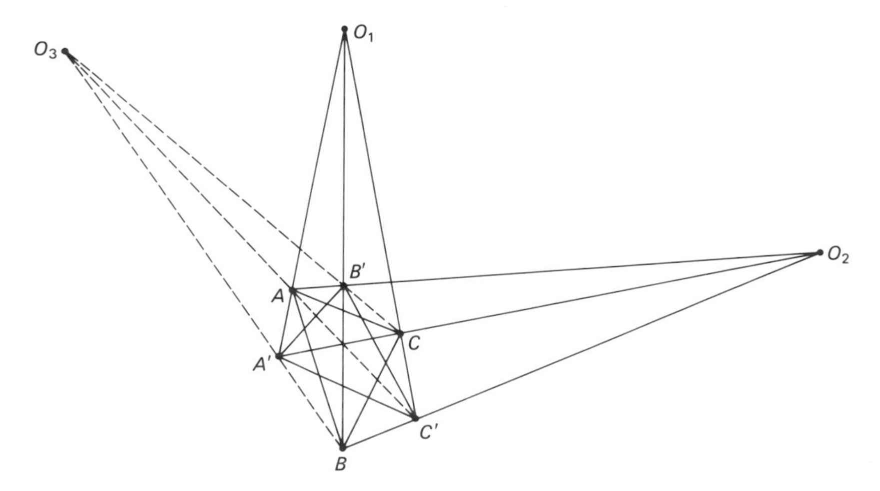

Figure 5.12

COORDINATIZATION

CHAPTER 6

In Chapter 3, Basis 3.5 was shown to be a projective plane; in Theorem 5.5., to be desarguesian; and, in Theorem 5.11, to be pappian. The commutativity of multiplication for real numbers was used only in the proof of Theorem 5.11. Bases 3.6–3.9 can be shown, in like manner, to be desarguesian and, with the use of commutativity, pappian projective planes. The algebraic properties of the real numbers used in the verification for Basis 3.5 hold also for the number systems in these other bases. It would seem, then, that algebraic incidence bases with elements from certain number systems are desarguesian (or even pappian) projective planes. Indeed, we shall see that this is the case.

Now we can ask whether every desarguesian (or pappian) projective plane is obtainable as such an algebraic incidence basis. This is analogous to the question posed in Chapter 4. Can every affine plane be obtained as a deletion subgeometry of some projective plane? It followed, there, that the intrinsic properties (specifically, those making it an affine plane) of a deletion subgeometry were sufficient to give an affirmative answer.

We seek, also, an affirmative answer to our present inquiry. The property that these algebraic incidence bases are desarguesian projective planes will be seen to be sufficient to guarantee that any desarguesian projective plane be so obtainable. Stated more precisely, any desarguesian projective plane is isomorphic to an algebraic incidence basis with elements from some particular number system. We shall work for a description of these number systems, which will give us, then, the abstract bond among the five number systems used in Bases 3.5–3.9. Further, we shall see how the pappian property corresponds to the commutativity of multiplication in the number systems.

We should mention, at this time, the more general question of obtaining any projective plane as an algebraic incidence basis. There is a sense in which this can be done but the problem is much more general, in fact, too general to be considered at this time. (This question is discussed by Marshall Hall, Jr. [3].)

In each of the next three sections an introductory motivation is given in terms of euclidean geometry. This is not to be confused with the projective theory!

6.1. COORDINATES

We shall look at an alternate and equivalent way of introducing coordinates in ordinary analytic geometry. Let there be given, in a euclidean plane, an ordered pair (x, y) of perpendicular lines (see Figure 6.1). Further, let Z be their point of intersection, l a line on Z different from x and y, and U a point on l different from Z. Now consider the number scale on l using Z as the origin and U as the unit point (that is, Z and U would be designated by 0 and 1, respectively, in this number scale). For any point P, let a be the real number designating in this number scale the intersection of l and the line on P parallel or equal to y, and b be the real number designating the intersection of l and the line on P parallel or equal to x. Then P is defined to have coordinates consisting of the ordered pair of real numbers (a, b). We can see readily that P would have the same coordinates (a, b) determined in the usual way relative to a cartesian coordinate system with x and y as axes if the unit points on x and y were, respectively, the points with coordinates $(1, 0)$ and $(0, 1)$ determined by the present scheme. Note that the point on l which would be designated by the real number a (in the number scale on l) has coordinates (a, a).

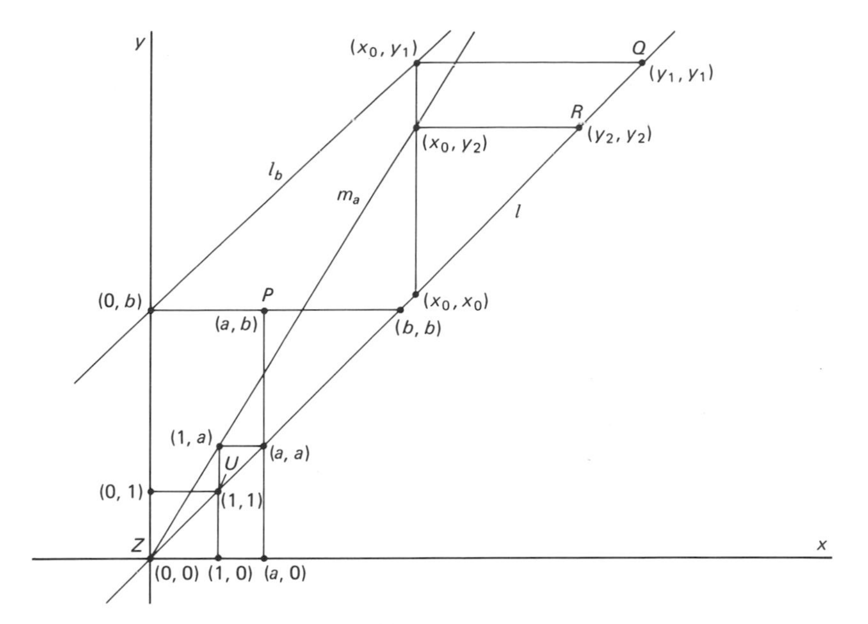

Figure 6.1

In addition, the number scale on l is, in essence, a one-to-one correspondence between the set of points on l and the set of real numbers. Finally, to facilitate the emulation of the above construction in a projective plane, consider the given euclidean plane to be imbedded into a projective plane. Then the projective points $L(Z)$, $L(U)$, $L(x)$, and $L(y)$ are distinct and no three are collinear. They can be used also to describe the projective lines $\{x\}$, $\{y\}$, and $\{l\}$, and the above construction of coordinates when interpreted in the imbedding. This discussion in a euclidean plane is the motivating basis for what follows.

In the remainder of this section $(\mathscr{P}, \mathscr{L}, \circ)$ is to be a projective plane with join $\vee$ and intersection $\wedge$.

An ordered quadruple (Z, U, J, K) of distinct points such that no three are collinear will be called a *coordinate system*. Let (Z, U, J, K) be a coordinate system and define the following elements:

$$l = Z \vee U, \qquad j = Z \vee J, \qquad k = Z \vee K, \qquad m = J \vee K, \qquad I = l \wedge m$$

(see Figure 6.2). We shall refer to them as the *auxiliary elements associated with* (Z, U, J, K). Now define

$$\mathscr{F} = \{P \in \mathscr{P} : P \circ l \quad \text{and} \quad P \neq I\}.$$

Finally, for any point $P \varnothing m$, define $X = l \wedge (P \vee K)$, $Y = l \wedge (P \vee J)$ and the ordered pair (X, Y) to be the *coordinates* of P, written $P: (X, Y)$.

THEOREM 6.1. There exists a one-to-one correspondence between $\mathscr{P}' = \{P \in \mathscr{P} : P \varnothing m\}$ and $\mathscr{F} \times \mathscr{F}$.

The proof is left as an exercise.

EXERCISE

6.1. Prove Theorem 6.1.

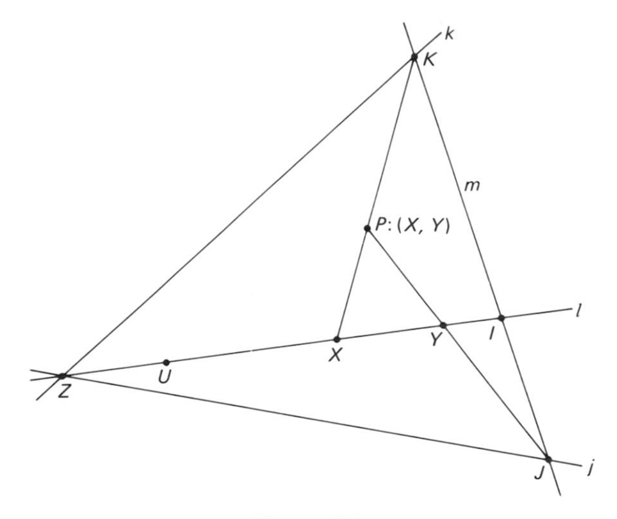

Figure 6.2

6.2. ADDITION

We shall, again, use a euclidean motivation to introduce the next concept. Assume the hypotheses of Figure 6.1. Given the real numbers x_0 and b (hence the points on l designated by them), we should like to find the point on l designated by $x_0 + b$. To this end, consider the line l_b with equation $y = x + b$, which can be described geometrically as the line on $(0, b)$ parallel to l. Now there is a point on l_b with abscissa x_0, in particular, the intersection of l_b and the line on (x_0, x_0) parallel or equal to y. Further, if (x_0, y_1) is on l_b, then $y_1 = x_0 + b$. Thus the point Q on l designated by $x_0 + b$ is the point with coordinates (y_1, y_1) where (x_0, y_1) is on l_b, that is, Q is the intersection of l and the line on (x_0, y_1) parallel or equal to x. We have, in effect, geometrically described "the process of adding b to x_0." More generally, the line l_b is the key to describing "the process of adding b to other real numbers." Stated another way, l_b is the key to describing the function whose value is $x' + b$ for every real number x'.

In the remainder of this section $(\mathscr{P}, \mathscr{L}, \circ)$ is to be a desarguesian projective plane. Let (Z, U, J, K) be a coordinate system with associated auxiliary elements l, j, k, m, I and the set $\mathscr{F}$ as defined in the last section.

Let $B \in \mathscr{F}$. Define $l_B = ((J \vee B) \wedge k) \vee I$. In addition, define α_B, a function from $\mathscr{F}$ to the set of points on l, by

$$\alpha_B A = (((K \vee A) \wedge l_B) \vee J) \wedge l \qquad \text{for} \quad A \in \mathscr{F}$$

(see Figure 6.3).

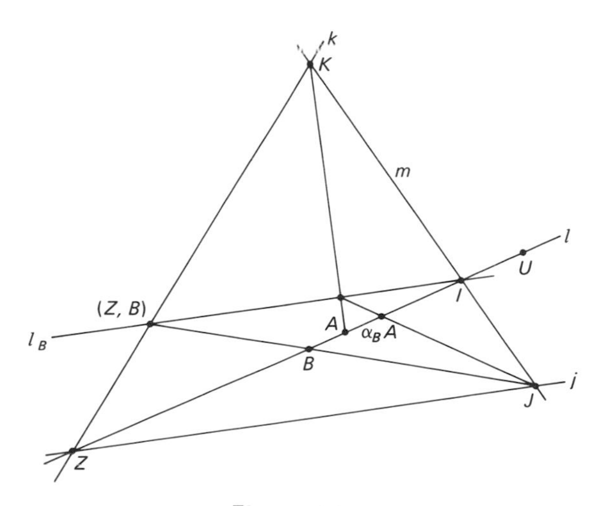

Figure 6.3

COROLLARY 6.2.

(a) l_B is a well-defined line;

(b) α_B is a function on $\mathscr{F}$ to $\mathscr{F}$.

PROOF. Let $B \in \mathscr{F}$. For part (a) we must show that the joins and intersections in the defining expression for l_B are formed for distinct pairs of points and lines, respectively. For part (b), let $A \in \mathscr{F}$. Now we must show that $\alpha_B A$ is a well-defined point on l different from I. The details are left as an exercise.

Define $+$ by $A + B = \alpha_B A$ for $A, B \in \mathscr{F}$. Then $+$, called *addition*, is a binary operation on $\mathscr{F} \times \mathscr{F}$ to $\mathscr{F}$.

COROLLARY 6.3. Let $B \in \mathscr{F}$, $l_B = ((J \vee B) \wedge k) \vee I$, and $P \not\circ m$ such that P: (X, Y). Then $P \circ l_B$ if and only if $Y = X + B$.

We shall say that $Y = X + B$ is an *equation* of the line l_B.

PROOF OF COROLLARY 6.3. Let $B \in \mathscr{F}$, $l_B = ((J \vee B) \wedge k) \vee I$, and $P \not\circ m$ such that P:(X, Y). Then $P \circ K \vee X, J \vee Y$. For the "only if" part, let $P \circ l_B$. Then

$$X + B = (((K \vee X) \wedge l_B) \vee J) \wedge l = (P \vee J) \wedge l = Y.$$

For the converse, let $Y = X + B$. Define $Q = (K \vee X) \wedge l_B$. Then $Y = (Q \vee J) \wedge l$. Now $P, Q \circ J \vee Y$. But $P, Q \circ K \vee X$. Thus $P = Q$, so that $P \circ l_B$.

THEOREM 6.4.

(a) For $A, B, C \in \mathscr{F}$, $(A + B) + C = A + (B + C)$.

(b) For $A \in \mathscr{F}$, $A + Z = A$ and $Z + A = A$. Moreover, Z, with this property, is unique.

(c) For $A \in \mathscr{F}$, there exists unique $B \in \mathscr{F}$ such that $A + B = Z$ and $B + A = Z$.

(d) For $A, B \in \mathscr{F}$, $A + B = B + A$.

The properties expressed in Theorem 6.4 are used, respectively, to make the following definitions:

(a) $+$ is said to be *associative*;
(b) Z will be called the *additive identity*;
(c) B will be called the *additive inverse of* A and written $-A$;
(d) $+$ is said to be *commutative*.

PROOF OF THEOREM 6.4. Part (b) is left as an exercise. For part (d), let $A, B \in \mathscr{F}$. If $A = Z$, $B = Z$, or $A = B$, the result is immediate. Let $A, B \neq Z$ and $A \neq B$. Define $A' = (J \vee A) \wedge k$, $B' = (J \vee B) \wedge k$, $l_A = I \vee A'$, $l_B = l \vee B'$, $A_1 = (K \vee A) \wedge l_A$, $B_1 = (K \vee B) \wedge l_A$, $A_2 = (K \vee A) \wedge l_B$, and $B_2 = (K \vee B) \wedge l_B$ (see Figure 6.4). Then $A_2 : (A, A + B)$ and $B_1 : (B, B + A)$. In order that $A + B = B + A$, we need A_2, B_1, and J collinear. As suggested by the conclusion of the Theorem

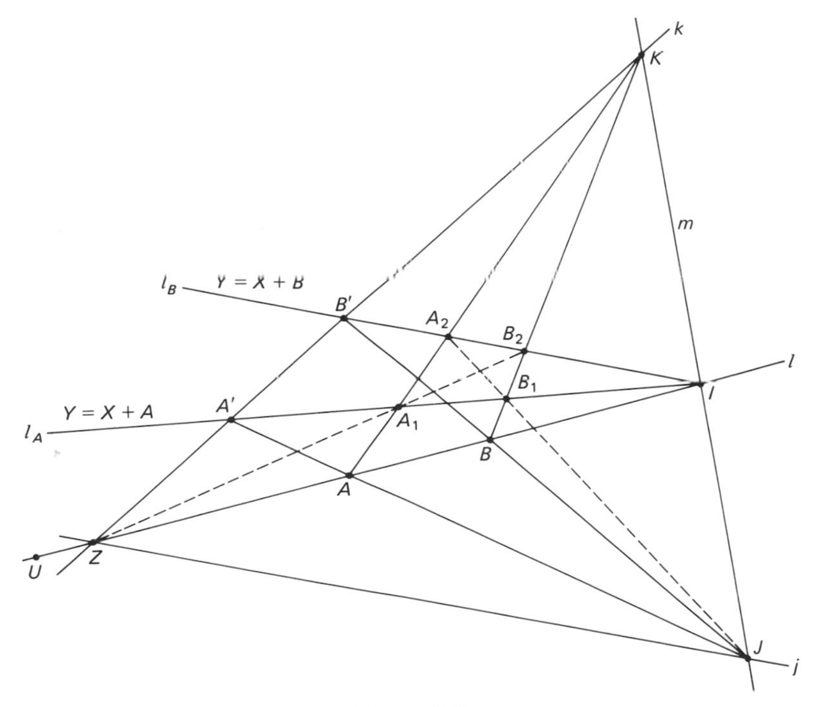

Figure 6.4

of Desargues we might look for triangles for which the intersections of corresponding sides are A_2, B_1, and J. Then if the triangles are perspective from a point we can apply the Theorem of Desargues to obtain the necessary collinearity. When looking for such triangles, it will be advisable, also, to confine ourselves, insofar as it is possible, to points and lines already present in our configuration and not to introduce additional joins and intersections. For example, we might consider triangles $AA'A_1$ and $B'BB_2$ for which we have

$$(A \vee A_1) \wedge (B' \vee B_2) = A_2,$$
$$(A' \vee A_1) \wedge (B \vee B_2) = B_1,$$
$$(A \vee A') \wedge (B' \vee B) = J.$$

However, in order to determine if they are perspective from a point, we should have to introduce $A \vee B'$, $A' \vee B$ and $A_1 \vee B_2$ into our configuration. Let us consider, instead, triangles $BB'B_2$ and IKA_1. Now $A_1 \vee B_2$ is the only join introduced; and in order that the triangles be perspective from a point, we need $A_1 \vee B_2 \circ (B \vee I) \wedge (B' \vee K) = Z$. Thus our need is shifted to the collinearity of Z, A_1, and B_2. As before we can look for triangles to be used in the Theorem of Desargues to establish this collinearity. For example, the corresponding sides of $A'B'I$ and ABK intersect appropriately and these triangles are perspective from J. Now we shall reverse the above analysis and present our argument in the following concise and deductive form. We have triangles $A'B'I$ and ABK perspective from J. Thus, by the Theorem of Desargues, $(A' \vee B') \wedge (A \vee B) = Z$, $(A' \vee I) \wedge (A \vee K) = A_1$ and $(B' \vee I) \wedge (B \vee K) = B_2$ are collinear. Hence triangles $BB'B_2$ and IKA_1 are perspective from Z, so that by the Theorem of Desargues, A_2, B_1, and J are collinear. Now $A + B = (A_2 \vee J) \wedge l = (B_1 \vee J) \wedge l = B + A$. This completes the proof of part (d).

For part (a), let $A, B, C \in \mathscr{F}$. Now let $A, B, C \neq Z$ and $A \neq C$. (The contrary special cases are left as an exercise.) Define $B' = (J \vee B) \wedge k$, $C' = (J \vee C) \wedge k$, $l_B = I \vee B'$, $l_C = I \vee C'$, $A_1 = (K \vee A) \wedge l_B$, $D = (J \vee A_1) \wedge l$, $D_1 = (K \vee D) \wedge l_C$, $B_1 = (K \vee B) \wedge l_C$, $E = (J \vee B_1) \wedge l$, $E' = (J \vee E) \wedge k$, $l_E = I \vee E'$, $A_2 = (K \vee A) \wedge l_E$, $F = (A_1 \vee D_1) \wedge m$ (Figure 6.5). Then $D = A + B$, $E = B + C$, $D_1 : (A + B, (A + B) + C)$ and $A_2 : (A, A + (B + C))$. Now we need A_2, D_1, and J collinear.

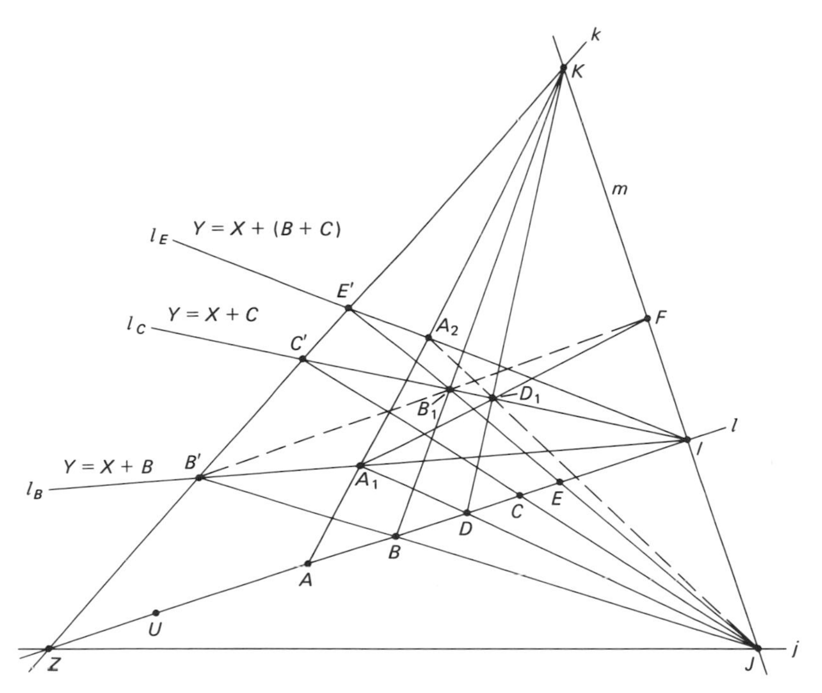

Figure 6.5

To this end, we have triangles BJK and IA_1D_1 perspective from D. Thus $(B \vee J) \wedge (I \vee A_1) = B'$, $(J \vee K) \wedge (A_1 \vee D_1) = F$, and $(B \vee K) \wedge (I \vee D_1) = B_1$ are collinear by the Theorem of Desargues. Now triangles $E'IB_1$ and KA_1F are perspective from B', so that A_2, D_1, and J are collinear. Thus $(A + B) + C = (D_1 \vee J) \wedge l = (A_2 \vee J) \wedge l = A + (B + C)$. This completes the proof of part (a).

Finally, for part (c), let $A \in \mathscr{F}$. In case $A = Z$, we have $Z + Z = Z$. Let $A \neq Z$. First, we shall find $B \in \mathscr{F}$ such that $B + A = Z$. To this end, let $l_A = ((J \vee A) \wedge k) \vee I$ (see Figure 6.6). Now find the point B_1 on l_A with second coordinate equal to Z, namely, $B_1 = l_A \wedge j$. Define $B = (K \vee B_1) \wedge l$. Then $B_1 : (B, Z)$, so that $Z = B + A$. Next, we have, from part (d), $A + B = B + A = Z$. Finally, for the uniqueness of B, let $B' \in \mathscr{F}$ such that $A + B' = Z$ and $B' + A = Z$. Then $B' = B' + Z = B' + (A + B) = (B' + A) + B = Z + B = B$. This completes the proof of Theorem 6.4.

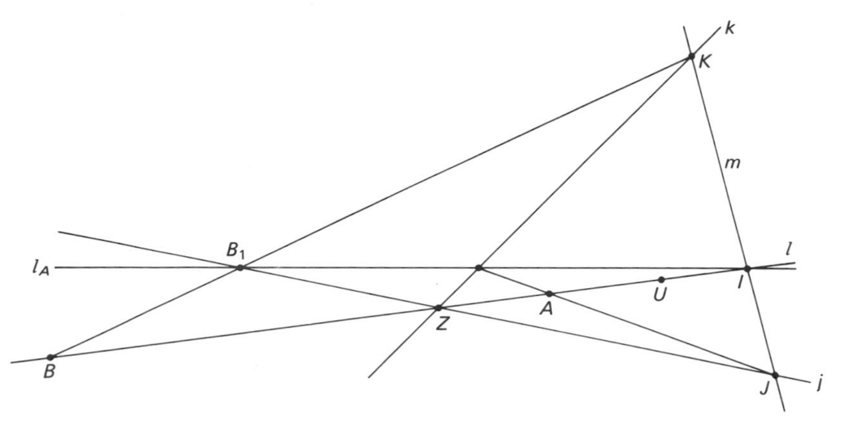

Figure 6.6

EXERCISES

6.2. Complete the proof of Corollary 6.2.

6.3. Prove part (b) of Theorem 6.4.

6.4. Complete the proof of part (a) of Theorem 6.4.

6.5. In Figure 6.7, consider the geometry of the paper as an affine plane imbedded into a desarguesian projective plane. Find A_i', B on l such that $L(A_i') = -L(A_i)$ for $i = 1, 2, 3, 4$ and $L(B) = -L(l)$.

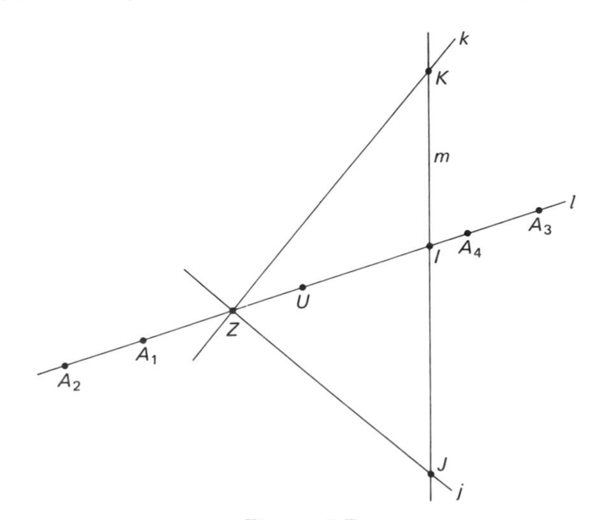

Figure 6.7

6.3. MULTIPLICATION

We shall return to the hypotheses of Figure 6.1. Given the real numbers x_0 and a such that $a \neq 0$ (hence the points on l designated by them), we should like to find the point on l designated by $a \cdot x_0$. Let us consider the line m_a with equation $y = a \cdot x$, that is, the line which is the join of Z and $(1, a)$. Then the point on m_a with abscissa x_0 has coordinates (x_0, y_2), where $y_2 = a \cdot x_0$. Hence the point R on l with coordinates (y_2, y_2) is designated by $a \cdot x_0$ in the number scale on l. Thus we have a geometric description of "the process of multiplying a by x_0." Similar to the role l_b played for addition of real numbers, m_a is the key to describing the function whose value is $a \cdot x'$ for every real number x'. This euclidean discussion provides the basis for what follows.

In the remainder of this section $(\mathscr{P}, \mathscr{L}, \circ)$ is to be a desarguesian projective plane. Let (Z, U, J, K), l, j, k, m, I, and $\mathscr{F}$ be as defined in the last section.

Let $A \in \mathscr{F}$. Define $m_A = ((J \vee A) \wedge (K \vee U)) \vee Z$ (see Figure 6.8). In addition, define μ_A, a function from $\mathscr{F}$ to the set of points on l, by

$$\mu_A B = (((K \vee B) \wedge m_A) \vee J) \wedge l \qquad \text{for} \quad B \in \mathscr{F}.$$

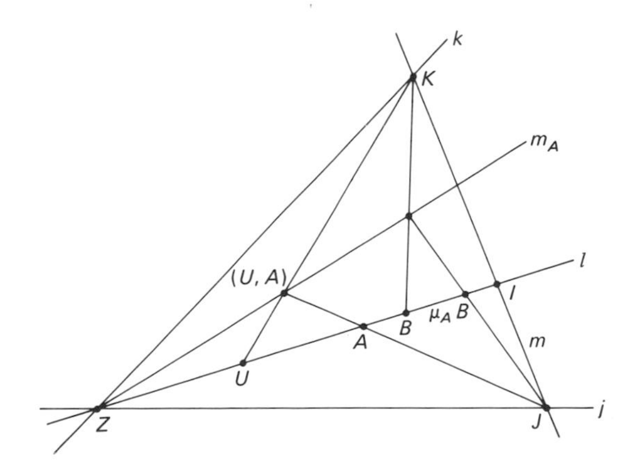

Figure 6.8

COROLLARY 6.5. For $A \in \mathscr{F}$.

(a) m_A is a well-defined line;
(b) μ_A is a function on $\mathscr{F}$ to $\mathscr{F}$.

The proof is left as an exercise.

Define $\cdot$ by $A \cdot B = \mu_A B$ for $A, B \in \mathscr{F}$. Then $\cdot$, called *multiplication*, is a binary operation on $\mathscr{F} \times \mathscr{F}$ to $\mathscr{F}$.

COROLLARY 6.6 Let $A \in \mathscr{F}$, $m_A = ((J \vee A) \wedge (K \vee U)) \vee Z$, and $P \not o m$ such that $P : (X, Y)$. Then $P \circ m_A$ if and only if $Y = A \cdot X$.

We shall say that $Y = A \cdot X$ is an *equation* of the line m_A.

COROLLARY 6.7. Let $A, B \in \mathscr{F}$. Then $A \cdot B = Z$ if and only if $A = Z$ or $B = Z$.

The proofs of these corollaries are left as exercises.

THEOREM 6.8.

(a) For $A, B, C \in \mathscr{F}$, $(A \cdot B) \cdot C = A \cdot (B \cdot C)$.
(b) For $A \in \mathscr{F}$, $A \cdot U = A$ and $U \cdot A = A$. Moreover, U, with this property, is unique.
(c) For $A \in \mathscr{F}$ such that $A \neq Z$, there exists unique $B \in \mathscr{F}$ such that $A \cdot B = U$ and $B \cdot A = U$.
(d) For $A, B, C \in \mathscr{F}$, $A \cdot (B + C) = A \cdot B + A \cdot C$ and $(A + B) \cdot C = A \cdot C + B \cdot C$.

The properties expressed in Theorem 6.8 are used, respectively, to make the following definitions:

(a) $\cdot$ is said to be *associative*;
(b) U will be called the *multiplicative identity*;
(c) B will be called the *multiplicative inverse* of A and written A^{-1};
(d) $\cdot$ is said to be *distributive* (*left* and *right*, respectively) *with respect to* $+$.

PROOF OF THEOREM 6.8. Part (b) is left as an exercise. For part (a), let $A, B, C \in \mathcal{F}$. The case that one of A, B, or C equals Z or U is left as an exercise. Let $A, B, C \neq Z, U$. Define $B' = (J \vee B) \wedge (K \vee U)$, B_1 such that $B_1 : (B, A \cdot B)$, C_1 such that $C_1 : (C, B \cdot C)$, $D = A \cdot B$, $D' = (J \vee D) \wedge (K \vee U)$, C_2 such that $C_2 : (C, (A \cdot B) \cdot C)$, $E = B \cdot C$, E_1 such that $E_1 : (B \cdot C, A \cdot (B \cdot C))$, and $F = (C_1 \vee D') \wedge (B_1 \vee E)$ (see Figure 6.9). We need C_2, E_1, and J collinear. We have triangles BB_1E

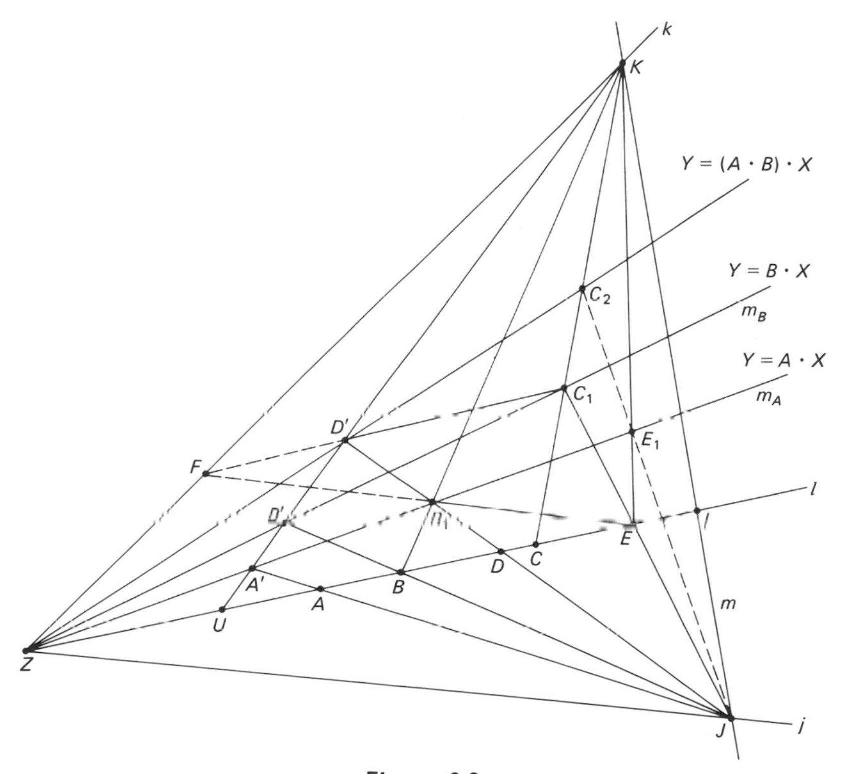

Figure 6.9

and $B'D'C_1$ perspective from J, so that by the Theorem of Desargues, we have the collinearity of K, F, and Z. Hence $ZD'B_1$ and KC_1E are perspective from F. Therefore C_2, J, and E_1 are collinear. Now $(A \cdot B) \cdot C = (J \vee C_2) \wedge l = (J \vee E_1) \wedge l = A \cdot (B \cdot C)$.

For part (c), let $A \in \mathscr{F}$ such that $A \neq Z$. In case $A = U$, we have $U \cdot U = U$. Let $A \neq U$. First, find $B \in \mathscr{F}$ such that $A \cdot B = U$. To this end, let A' be such that $A' : (U, A)$ and $m_A = Z \vee A'$. Find the point B_1 on m_A with second coordinate U; namely, $B_1 = (J \vee U) \wedge m_A$. Define $B = (K \vee B_1) \wedge l$. Then $B_1 : (B, U)$, so that $U = A \cdot B$ since $B_1 \circ m_A$. Now we must show that $B \cdot A = U$ and that B is unique. These are left as an exercise.

For part (d), let $A, B, C \in \mathscr{F}$. We shall prove the right distributivity first. For this purpose, let $A, B \neq Z$ and $C \neq Z, U$. (The remaining cases are left as an exercise.) Define $A_1 = (K \vee A) \wedge (J \vee B)$, $B' = (J \vee B) \wedge k$, D such that $D : (A, A + B)$, $F = A \cdot C$, E such that $E : (C, B \cdot C)$, E' such that $E' : (Z, B \cdot C)$, $F_1 = (K \vee F) \wedge (J \vee E)$ and $F_2 = (K \vee F) \wedge (I \vee E')$ (Figure 6.10). Then $Z \vee A_1$ has equation $Y = (B \cdot A^{-1}) \cdot X$ since $A_1 : (A, B)$ and $(B \cdot A^{-1}) \cdot A = B \cdot (A^{-1} \cdot A) =$

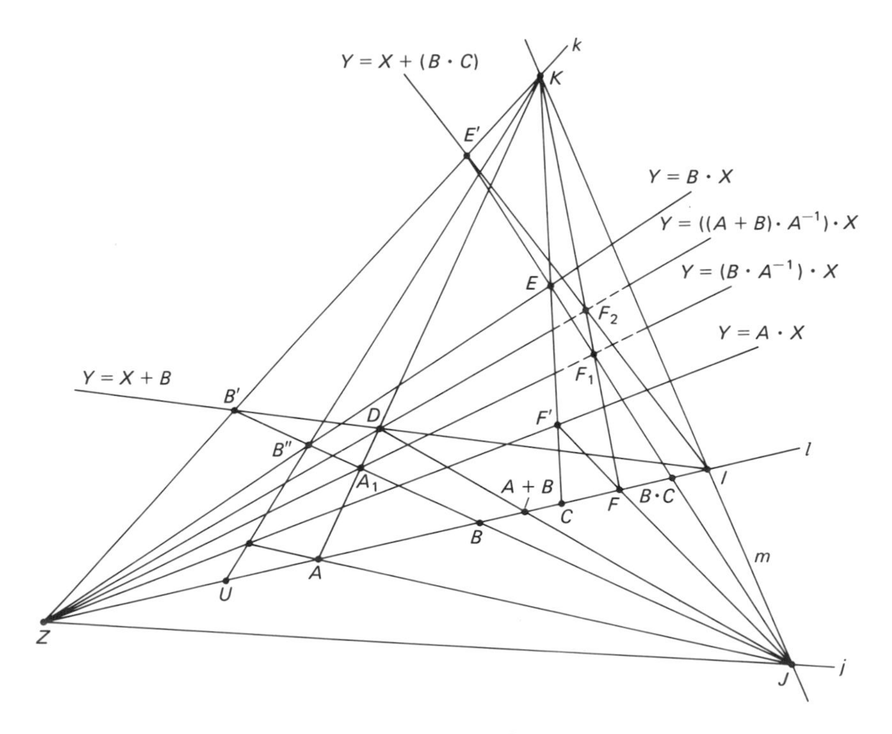

Figure 6.10

$B \cdot U = B$. Further, $F_1 \circ Z \vee A_1$ since $F_1 : (A \cdot C, B \cdot C)$ and $(B \cdot A^{-1}) \cdot (A \cdot C) = ((B \cdot A^{-1}) \cdot A) \cdot C = (B \cdot (A^{-1} \cdot A)) \cdot C = (B \cdot U) \cdot C = B \cdot C$. Now triangles $B'A_1A$ and $E'F_1F$ are perspective from Z, whence from m. Define $O = (B' \vee A) \wedge (E' \vee F)$. Then $O \circ m$. We also have triangles $B'IE'$ and AKF perspective from O, so that D, F_2, and Z are collinear. Now $Z \vee D$ has equation $Y = ((A + B) \cdot A^{-1}) \cdot X$ since $D: (A, A + B)$ and

$$((A + B) \cdot A^{-1}) \cdot A = (A + B) \cdot (A^{-1} \cdot A) = (A + B) \cdot U = A + B.$$

Hence since $F_2 \circ Z \vee D$ and F_2 has first coordinate $A \cdot C$, F_2 has second coordinate

$$\begin{aligned} ((A + B) \cdot A^{-1}) \cdot (A \cdot C) &= (A + B) \cdot (A^{-1} \cdot (A \cdot C)) \\ &= (A + B) \cdot ((A^{-1} \cdot A) \cdot C) \\ &= (A + B) \cdot (U \cdot C) \\ &= (A + B) \cdot C. \end{aligned}$$

But $F_2 : (A \cdot C, A \cdot C + B \cdot C)$ since $F_2 \circ I \vee E'$ and $I \vee E'$ has equation $Y = X + (B \cdot C)$. Thus, by Theorem 6.1, $(A + B) \cdot C = A \cdot C + B \cdot C$.

For the left distributivity in part (d), we still have $A, B, C \in \mathscr{F}$; however, let $A \neq Z, U$ and $B, C \neq Z$. (The remaining cases are left as an exercise.) Define $C' = (J \vee C) \wedge k$, B_1 such that $B_1 : (B, B + C)$, B_2 such that $B_2 : (B, A \cdot B)$, $D = B + C$, $E = A \cdot B$, C_1 such that $C_1 : (C, A \cdot C)$, $F = A \cdot C$, F' such that $F' : (Z, A \cdot C)$, $H = (Z \vee B_2) \wedge m$, $Q = (F' \vee H) \wedge (K \vee B)$, $P = (C' \vee H) \wedge (K \vee B)$, $G = (B_2 \vee I) \wedge (F \wedge H)$, $D_1 = (B_2 \vee I) \wedge (C \vee H)$, E_1 such that $E_1 : (A \cdot B, A \cdot B + A \cdot C)$ and D_2 such that $D_2 : (B + C, A \cdot (B + C))$ (see Figure 6.11). We shall show that D_2, E_1, and J are collinear to obtain $A \cdot (B + C) = A \cdot B + A \cdot C$. First, we have triangles KIB_2 and $C'CH$ perspective from Z, so that J, D_1, and P are collinear. Now triangles PJB_1 and HCI are perspective from C', and therefore D_1, D, and K are collinear. Hence $D_2 \circ K \vee D = K \vee D_1$. Now triangles KIB_2 and $F'FH$ are perspective from Z. Hence J, G, and Q are collinear. We also have triangles KEB_2 and $F'IH$ perspective from Z, so that E_1, J, and Q are collinear. From this, we have E_1, G, and J are collinear. Finally, we have triangles

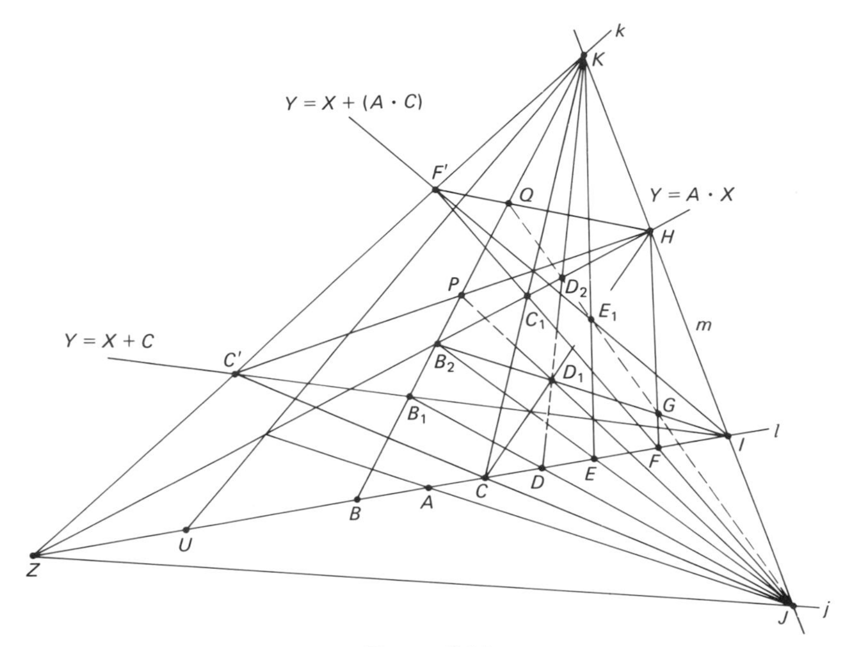

Figure 6.11

C_1HF and KD_1I perspective from C, and therefore D_2, G, J are collinear. Hence D_2, E_1, and J are collinear. Thus, by Theorem 6.1, $A \cdot (B + C) = A \cdot B + A \cdot C$. This completes the proof of Theorem 6.8.

We note that l has equations $Y = X + Z$ and $Y = U \cdot X$. However, in light of the algebraic properties of Z and U, these equations are compatible; in fact, both can be written as $Y = X$. In the next two theorems we extend the class of lines having an equation to include all lines different from m.

THEOREM 6.9. Let $A, B \in \mathscr{F}$ and let $l_B = ((J \vee B) \wedge k) \vee I$, $m_A = ((J \vee A) \wedge (K \vee U)) \vee Z$, $n = (m \wedge m_A) \vee (k \wedge l_B)$, and $P \varnothing m$ such that $P : (X, Y)$. Then $P \circ n$ if and only if $Y = A \cdot X + B$.

We shall say that $Y = A \cdot X + B$ is an *equation* of the line n.

PROOF OF THEOREM 6.9. Let A, B, l_B, m_A, n, P, X, and Y be as in the hypothesis of the theorem. In case $A = U$, we have $m_A = l$ and $n = l_B$. Hence n has equation $Y = X + B$, which can be written as $Y = U \cdot X + B$ in light of the multiplicative properties of U. In case $B = Z$, $l_B = l$, and $n = m_A$. Thus n has equation $Y = A \cdot X$, which can be written as $Y = A \cdot X + Z$. Now let $A \neq U$ and $B \neq Z$ (see Figure 6.12). We have, from the original hypothesis, $P \circ K \vee X$ and $P \circ J \vee Y$.

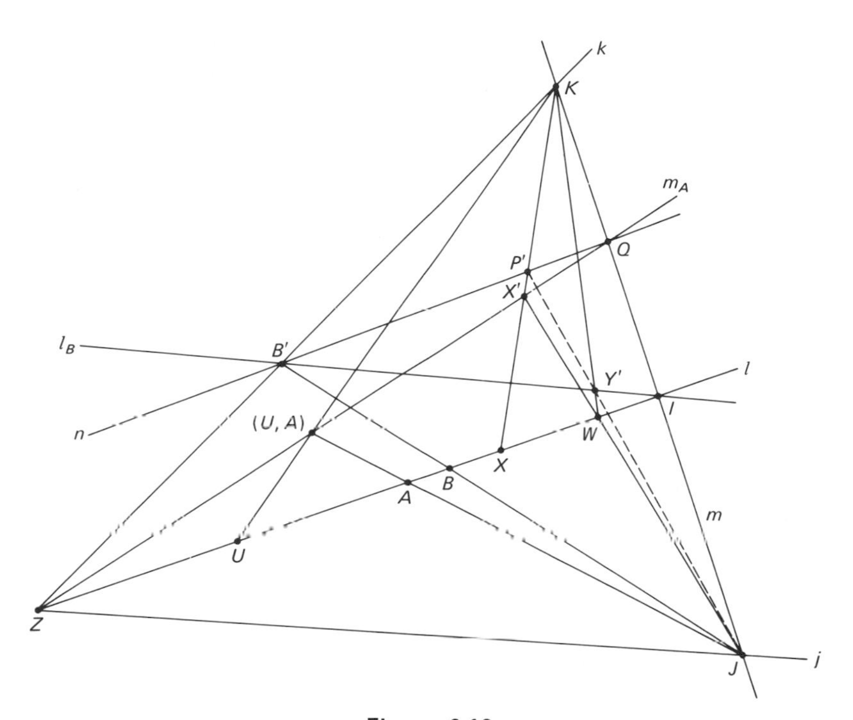

Figure 6.12

Define X' such that $X' : (X, A \cdot X)$, B' such that $B' : (Z, B)$, $W = A \cdot X$, Y' such that $Y' : (A \cdot X, A \cdot X + B)$, $Q = m \wedge m_A$, and $P' = (K \vee X) \wedge n$. Then $P' \circ n$. We have triangles $P'B'Y'$ and $X'ZW$ perspective from K. Hence Q, I, and $(P' \vee Y') \wedge (X' \vee W)$ are collinear. Therefore $P' \vee Y' \circ (Q \vee I) \wedge (X' \vee W) = J$. Thus $P' : (X, A \cdot X + B)$. Now for the "only if" part, let $P \circ n$. Then $P = (K \vee X) \wedge n = P'$.

Hence $P' : (X, Y)$. Thus, by Theorem 6.1, $Y = A \cdot X + B$. For the converse, let $Y = A \cdot X + B$. Then $P' : (X, Y)$. Hence $P = P'$ by Theorem 6.1. Thus $P \circ n$.

COROLLARY 6.10. For $n \varnothing K$, there exist unique $A, B \in \mathscr{F}$ such that n has equation $Y = A \cdot X + B$ and, conversely, for $A, B \in \mathscr{F}$, there exists a unique line not on K with equation $Y = A \cdot X + B$.

The proof is left as an exercise.

THEOREM 6.11. Let $n \circ K$ such that $n \neq m$, $C = n \wedge l$ and $P \varnothing m$ such that $P : (X, Y)$. Then $P \circ n$ if and only if $X = C$.

We shall say that $X = C$ is an *equation* of the line n. The proof of the theorem is left as an exercise.

THEOREM 6.12. If $(\mathscr{P}, \mathscr{L}, \circ)$ is pappian, $A \cdot B = B \cdot A$ for $A, B \in \mathscr{F}$.

We shall say that $\cdot$ is *commutative* when this property holds.

PROOF OF THEOREM 6.12. Let A, $B \in \mathscr{F}$. We shall consider only the case when $A, B \neq Z, U$ and $A \neq B$; the contrary cases follow from the algebraic properties already developed. Define $A' = (J \vee A) \wedge (K \vee U)$, $m_A = Z \vee A'$, $B' = (J \vee B) \wedge (K \vee U)$, $m_B = Z \vee B'$, $A_1 = (K \vee A) \wedge m_B$ and $B_1 = (K \vee B) \wedge m_A$ (see Figure 6.13). Consider the pappian point-cycle $\overline{A'ZB'BKA}$. Then by the Theorem of Pappus, $(A' \vee Z) \wedge (B \vee K) = B_1$, $(Z \vee B') \wedge (K \vee A) = A_1$, and $(B' \vee B) \wedge (A \vee A') = J$ are collinear. Thus $A \cdot B = (B_1 \vee J) \wedge l = (A_1 \vee J) \wedge l = B \cdot A$.

We conclude this section with the definition of two additional concepts to be used in later chapters. Define $-$, called *subtraction*, by $A - B = A + (-B)$ for $A, B \in \mathscr{F}$. Then $-$ is a binary operation on $\mathscr{F} \times \mathscr{F}$ to $\mathscr{F}$. When $(\mathscr{P}, \mathscr{L}, \circ)$ is pappian (so that $\cdot$ is commutative), define $\div$, called *division*, by $A \div B = A \cdot B^{-1}$ for $A, B \in \mathscr{F}$ such that $B \neq Z$. We shall also write A/B and $\dfrac{A}{B}$ for $A \div B$.

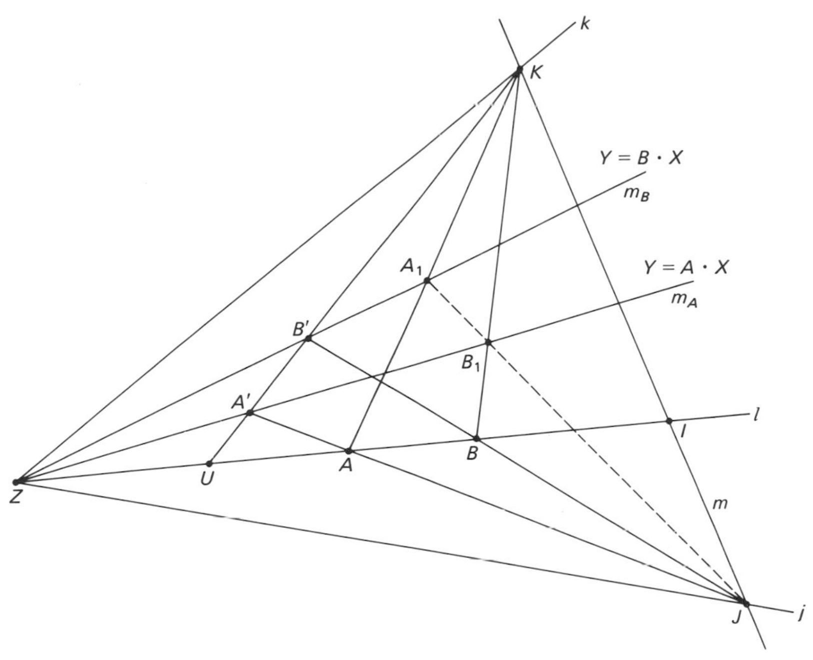

Figure 6.13

EXERCISES

6.6. Prove Corollary 6.5.
6.7. Prove Corollary 6.6.
6.8. Prove Corollary 6.7.
6.9. Prove part (b) of Theorem 6.8.
6.10. Prove $(-U) \cdot (-U) = U$.
6.11. Complete the proof of part (a) of Theorem 6.8.
6.12. For part (c) of Theorem 6.8, prove $B \cdot A = U$ and the uniqueness of B. Why is $A \neq Z$?
6.13. In Figure 6.14, consider the geometry of the paper as an affine plane imbedded into a desarguesian projective plane. Find A_i', B on l such that $L(A_i') = L(A_i)^{-1}$ for $i = 1, 2, 3, 4$ and $L(B) = L(l)^{-1}$.
6.14. Complete the proof of part (d) of Theorem 6.8.
6.15. Prove Corollary 6.10.

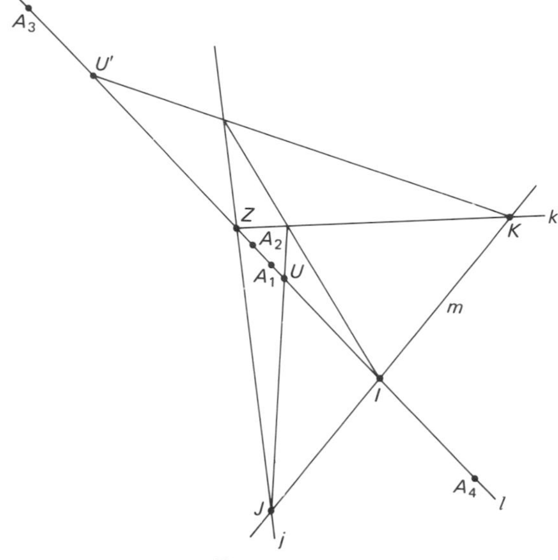

Figure 6.14

6.16. Prove Theorem 6.11.

6.17. Consider the projective plane obtained by imbedding the euclidean plane stated in the hypotheses of Figure 6.1. Select the coordinate system $(L(Z), L(U), L(x), L(y))$ in the projective plane and compare the results of Theorems 6.9 and 6.11 with the results of ordinary analytic geometry. In particular, the concepts of y-intercept and slope have what analogy in the projective situation?

6.18. In the proof of Theorem 6.12, show that $\overline{A'ZB'BKA}$ is a pappian point-cycle.

6.4. ALGEBRAIC SYSTEMS AND INCIDENCE BASES

Finally, we are in a position to describe fully those algebraic systems that we want to consider abstractly. Let $\mathscr{F}$ be a set and $+, \cdot$ binary operations on $\mathscr{F} \times \mathscr{F}$ to $\mathscr{F}$. The system $(\mathscr{F}, +, \cdot)$ is said to be a *division ring* with $+$ and $\cdot$ called, respectively, *addition* and *multiplication*, if the following are satisfied:

1. $+$ is associative;
2. there exists an additive identity 0;
3. each $x \in \mathscr{F}$ has an additive inverse;
4. $+$ is commutative;
5. $\cdot$ is associative;
6. there exists a multiplicative identity 1 different from 0;
7. each $x \in \mathscr{F}$ such that $x \neq 0$ has a multiplicative inverse;
8. $\cdot$ is distributive with respect to $+$.

If, in addition to the above, $\cdot$ is commutative, $(\mathscr{F}, +, \cdot)$ is said to be a *field*. It follows that there is a unique additive identity and a unique multiplicative identity, and that additive and multiplicative inverses are unique. For $x \in \mathscr{F}$, we shall write $-x$ for the additive inverse of x and x^{-1} for the multiplicative inverse of x (provided $x \neq 0$).

The real number system (also the number systems used in Bases 3.6–3.9) is a division ring and exactly the properties making it a division ring were used in the verification that Basis 3.5 is a desarguesian projective plane (see Appendix B). Further, the real number system is a field which was sufficient to show that Basis 3.5 is pappian. These comments are summarized in the next theorem and the proof for Basis 3.5 applies here.

THEOREM 6.13. An algebraic incidence basis with elements from a division ring is a desarguesian projective plane. Moreover, if the division ring is a field, the projective plane is pappian.

If we had begun our study with a knowledge of these algebraic concepts, Basis 3.5 would have been constructed directly with elements from an arbitrary division ring rather than the specific one used, the real number system. In this way we should have eliminated the need for the specific construction of Bases 3.6–3.9 since the number systems employed are division rings also. This illustrates one of the values in considering such systems abstractly.

6.5. THE COORDINATIZATION THEOREM

We now have the necessary algebraic materials to begin the discussion of the sufficiency of the desarguesian (or pappian) property as a guarantee that such a projective plane be obtainable (isomorphically) as an algebraic incidence basis with elements from some division ring (or field). This will be a form of converse to Theorem 6.13. The next theorem is a summary of the results obtained in the previous sections and restated in the present terminology.

THEOREM 6.14. Let (Z, U, J, K) be a coordinate system in a desarguesian projective plane $(\mathscr{P}, \mathscr{L}, \circ)$. Let $\mathscr{F}$, $+$, and $\cdot$ be defined as in Sections 6.2 and 6.3. Then $(\mathscr{F}, +, \cdot)$ is a division ring. Moreover, $(\mathscr{F}, +, \cdot)$ is a field if $(\mathscr{P}, \mathscr{L}, \circ)$ is pappian.

For (Z, U, J, K) a coordinate system in a projective plane, we shall refer to $(\mathscr{F}, +, \cdot)$ of the theorem as the *division ring* or *field associated with* (Z, U, J, K) according as $(\mathscr{P}, \mathscr{L}, \circ)$ is desarguesian or pappian. Actually, in a desarguesian projective plane, the division ring $(\mathscr{F}, +, \cdot)$ is fully determined by the specification of Z, U, I on l. This is, in essence, the content of the next theorem.

THEOREM 6.15. Let (Z, U, J, K) and (Z, U, J', K') be coordinate systems in a desarguesian projective plane such that $Z \vee U$, $J \vee K$, and $J' \vee K'$ are concurrent. Then the division rings $(\mathscr{F}, +, \cdot)$ and $(\mathscr{F}', +', \cdot')$ associated with (Z, U, J, K) and (Z, U, J', K'), respectively, are the same.

PROOF. First, note that the auxiliary elements l, I and l', I' for (Z, U, J, K) and (Z, U, J', K'), respectively, satisfy $l = l'$ and $I = I'$, whence $\mathscr{F} = \mathscr{F}'$. To show the equality of $+$ and $+'$, let $A, B \in \mathscr{F}$. If $A = Z$ or $B = Z$, we have $A + B = A +' B$ immediately. Let $A, B \neq Z$, (see Figure 6.15). Define $Q = (J \vee B) \wedge k$, $R = (Q \vee I) \wedge (K \vee A)$,

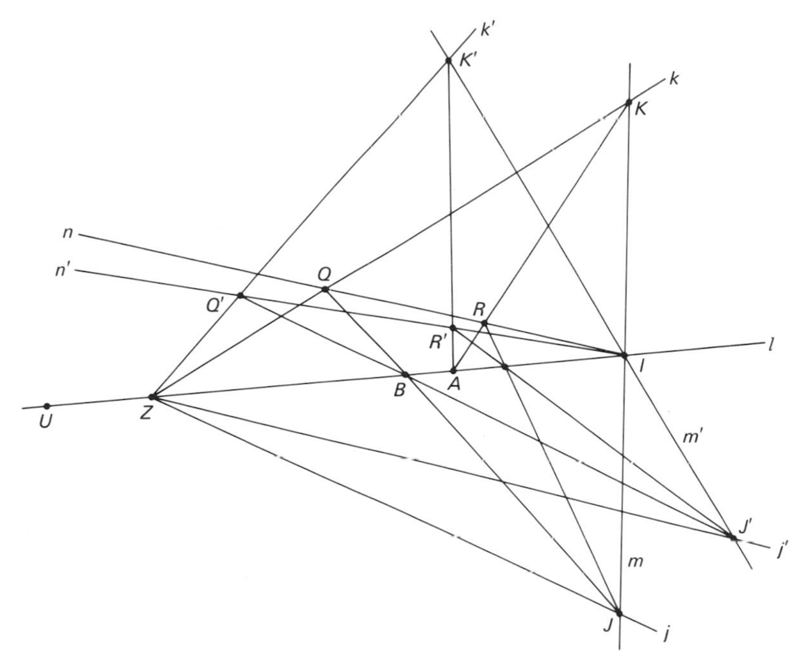

Figure 6.15

$Q' = (J' \vee B) \wedge k'$, and $R' = (Q' \vee I) \wedge (K' \vee A)$. Then $A + B = (I \vee R) \wedge l$ and $A +' B = (J' \vee R') \wedge l$. Now consider the complete quadrangles $KQRJ$ and $K'Q'J'R'$. We have

$$\begin{gathered}(K \vee Q) \wedge (K' \vee Q') = Z, \\ (K \vee J) \wedge (K' \vee J') = I, \\ (K \vee R) \wedge (K' \vee R') = A, \\ (Q \vee J) \wedge (Q' \vee J') = B, \\ (Q \vee R) \wedge (Q' \vee R') = I\end{gathered}$$

collinear on l, whence, by Theorem 5.6, $(J \vee R) \wedge (J' \vee R') \circ l$. Thus $A + B = A +' B$. The equality of $\cdot$ and $\cdot'$ is left as an exercise.

In view of the last theorem, we shall also refer to $(\mathscr{F}, +, \cdot)$ as the division ring (field) *associated with* (Z, U, I), where Z, U, I are distinct points on a line l in a desarguesian (pappian) projective plane. Now we consider the converse of the last part of Theorem 6.14.

THEOREM 6.16. If, in a desarguesian projective plane, the division ring associated with each choice of coordinate system is a field, the projective plane is pappian.

The proof is left as an exercise.

THEOREM 6.17. A desarguesian (pappian) projective plane is isomorphic to an algebraic incidence basis with elements from some division ring (field).

PROOF. It will be sufficient to prove the first part of the theorem. Let $(\mathscr{P}, \mathscr{L}, \circ)$ be a desarguesian projective plane. Let (Z, U, J, K) be a coordinate system with auxiliary elements l, j, k, m, and I, $(\mathscr{F}, +, \cdot)$ the division ring associated with (Z, U, J, K), and $(\mathscr{P}_1, \mathscr{L}_1, \circ_1)$ the algebraic incidence basis with elements from $(\mathscr{F}, +, \cdot)$. Define π and λ by the following. For $P \in \mathscr{P}$,

$$\pi P = \begin{cases} (S, T, U) & \text{if} \quad P \not\circ m \text{ and } P : (S, T), \\ (U, R, Z) & \text{if} \quad P \circ m, P \neq K, \text{and } P \text{ is on the line with equation } Y = R \cdot X, \\ (Z, U, Z) & \text{if} \quad P = K. \end{cases}$$

For $n \in \mathscr{L}$,

$$\lambda n = \begin{cases} [-A, U, -B] & \text{if} \quad n \not\circ K, \text{ and } n \text{ has equation } Y = A \cdot X + B, \\ [U, Z, -C] & \text{if} \quad n \circ K, n \neq m, \text{ and } n \text{ has equation } X = C, \\ [Z, Z, U] & \text{if} \quad n = m. \end{cases}$$

Certainly, π is a function with domain $\mathscr{P}$ and range contained in $\mathscr{P}_1$. To show that the range is all of $\mathscr{P}_1$, let $(X, Y, W) \in \mathscr{P}_1$. Then (X, Y, W) can be represented uniquely in exactly one of the forms (S, T, U), (U, R, Z), or (Z, U, Z). (These correspond, respectively, to the cases when $W \neq Z$, or $W = Z$ and $X \neq Z$, or $W = X = Z$ and $Y \neq Z$.) Hence we can find $P \in \mathscr{P}$ such that $\pi P = (X, Y, W)$. Thus $\mathscr{P}_1$ is the range of π. But the above argument also establishes that π is one-to-one for the point P we found is unique. Therefore π is a one-to-one correspondence between $\mathscr{P}$ and $\mathscr{P}_1$.

That λ is a one-to-one correspondence between $\mathscr{L}$ and $\mathscr{L}_1$ is left as an exercise.

Finally, to show that (π, λ) is an isomorphism, let $P \in \mathscr{P}$ and $n \in \mathscr{L}$ such that $P \circ n$. We shall consider three cases.

CASE 1. Let $n \not\circ K$. Then n has equation $Y = A \cdot X + B$ for unique $A, B \in \mathscr{F}$ and $\lambda n = [-A, U, -B]$. In case $P \not\circ m$, let $P : (S, T)$. Then $T = A \cdot S + B$. Also $\pi P = (S, T, U)$ and $\pi P \circ_1 \lambda n$ since $(-A) \cdot S + U \cdot T + (-B) \cdot U = -(A \cdot S + B) + T = -T + T = Z$. In case $P \circ m$, we have $P \neq K$. Thus P is on a line with equation $Y = R \cdot X$. Now $A = R$. Also $\pi P = (U, R, Z)$ and $\pi P \circ_1 \lambda n$ since $(-A) \cdot U + U \cdot R + (-B) \cdot Z = -A + R = -A + A = Z$.

CASE 2. Let $n \circ K$ such that $n \neq m$. Then $\lambda n = [U, Z, -C]$ where $C = n \wedge l$. In case $P \not\circ m$, let $P : (S, T)$. Then $S = C$. Also $\pi P = (S, T, U)$ and $\pi P \circ_1 \lambda n$ since

$$U \cdot S + Z \cdot T + (-C) \cdot U = S + (-C) = C + (-C) = Z.$$

In case $P \circ m$, we have $P = K$. Thus $\pi P = (Z, U, Z)$ and $\pi P \circ_1 \lambda n$ since $U \cdot Z + Z \cdot U + (-C) \cdot Z = Z$.

CASE 3. Let $n = m$. Then $\lambda n = [Z, Z, U]$. The remainder of this case is left as an exercise.

Now, by Theorem 4.1, (π, λ) is an isomorphism between $(\mathscr{P}, \mathscr{L}, \circ)$ and $(\mathscr{P}_1, \mathscr{L}_1, \circ_1)$. This completes the proof of Theorem 6.17.

The last result is our *Coordinatization Theorem*. We can think of the triples associated with the points of $\mathscr{P}$ under π as "coordinates" and the triples associated with the lines of $\mathscr{L}$ under λ as "coordinates." In fact, in many studies of projective planes, the coordinates for points that we introduced in Section 6.1 are called *nonhomogeneous coordinates*. Then the associated triples under π and λ mentioned above would be called *homogeneous coordinates*. We shall have no need for this additional terminology.

EXERCISES

6.19. Complete the proof of Theorem 6.15.
6.20. Prove Theorem 6.16.

6.21. Complete the proof of Theorem 6.17.

6.22. Let (Z, U, J, K) be a coordinate system in a fanian plane and let $(\mathscr{F}, +, \cdot)$ be its associated field. Describe $\mathscr{F}$, $+$, and $\cdot$. Further, describe π and λ of Theorem 6.17 for this special case (that is, *coordinatize* the fanian plane).

6.6. FINITE PROJECTIVE PLANES

A finite projective plane $(\mathscr{P}, \mathscr{L}, \circ)$ is one for which the sets $\mathscr{P}$ and $\mathscr{L}$ are finite. The set $\mathscr{F}$ associated with any coordinate system in a finite projective plane is necessarily finite; in fact, if $\mathscr{F}$ has n elements, there are exactly $n + 1$ points on each line. We shall call n the *order* of $(\mathscr{P}, \mathscr{L}, \circ)$. Now we state two theorems from abstract algebra which enable us to obtain a complete characterization of the finite desarguesian projective planes and the finite pappian projective planes.

THEOREM (Wedderburn). A finite division ring is a field.

THEOREM (Galois). If p is a prime integer and k a positive integer, there exists a unique field with $n = p^k$ elements, and conversely, the number of elements in any finite field is of this form.

These are very deep theorems from algebra, and no attempt will be made to prove them here or, even, to discuss them further. (We should mention that the uniqueness in the second theorem is in the sense of an isomorphism.) However, if we accept them and apply them to our theory, we have the following two theorems. The proofs are left as exercises.

THEOREM 6.18. A finite desarguesian projective plane is pappian.

THEOREM 6.19. If p is a prime integer, k a positive integer, and $n = p^k$, there exists a unique (isomorphically) desarguesian projective plane with exactly $n + 1$ points on a line and a total of $n^2 + n + 1$ points, and conversely, the number of points in a finite desarguesian projective plane is of this form.

Two immediate questions come to mind:

(a) whether there exist projective planes (necessarily nondesarguesian) with order n not of the form p^k;
(b) whether we can find nondesarguesian projective planes with order n of the form p^k, in fact, find all such planes.

Both of these questions have partial answers. Currently, they are central in a great deal of mathematical research activity.

It is known that a projective plane with order n *cannot* exist (shown in 1949 by Bruck and Ryser [7]) if n satisfies the following:

(a) $n = 4m + 1$ or $n = 4m + 2$ for some positive integer m;
(b) n is not the sum of two squares, that is, $n \neq a^2 + b^2$ for all nonnegative integers a and b.

The case $n = 6$ is the first not of the form p^k. In addition, 6 satisfies the above conditions, so that there is no projective plane of order 6. Actually, this result was known as early as 1900 (see reference [17]) in connection with another (and equivalent) problem. The next case not of the form p^k is $n = 10$. But 10 does not satisfy the above conditions, so that the statement sheds no light on the question of the existence of a projective plane of order 10. In fact, this case remains open in spite of a considerable amount of time spent on large computers in an attempt to construct a projective plane of this order. It can be shown that there are an infinite number of values of n excluded from being orders of finite projective planes by the above statement. On the other hand, it can also be shown that there are an infinite number of values of n which are neither of the form p^k nor excluded by the above statement. The existence of a projective plane for each of these orders remains an open question.

Relative to the second question, it is known that any finite projective plane of order $n \leqq 8$ is necessarily desarguesian. The existence of nondesarguesian projective planes of order $n = p^k$ with p odd and $k \geqq 2$ has been shown (see references [1] and [2]). There are also nondesarguesian planes of order $n = 2^{2k}$ with $k \geqq 2$. However, only the desarguesian plane is known for $n = p \geqq 11$.

EXERCISES

6.23. Prove Theorem 6.18.

6.24. Prove Theorem 6.19.

PROJECTIVITIES

CHAPTER 7

Functions (transformations) with domain and range consisting of points or lines play a fundamental role in many studies of geometry. We already have encountered such functions in our study, namely, isomorphisms. It is our purpose, in this chapter, to introduce some additional functions of this type and derive some properties to be used in the subsequent development.

Throughout this chapter we shall let $(\mathscr{P}, \mathscr{L}, \circ)$ be a projective plane.

7.1. PERSPECTIVITIES AND PROJECTIVITIES

Let $O \in \mathscr{P}$ and $l, m \in \mathscr{L}$ such that $l, m \not\circ O$. Define π by $\pi X = (X \vee O) \wedge m$ for $X \circ l$. We shall call π a *perspectivity between l and m with center O* and write $\pi: l \overset{O}{\overline{\overline{\wedge}}} m$. For π a perspectivity between l and m with center O and for $P_i \circ l$ and $Q_i = \pi P_i$ for $i = 1, \ldots, n$, we shall write $(P_1, \ldots, P_n) \overset{O}{\overline{\overline{\wedge}}} (Q_1, \ldots, Q_n)$ and say $(P_1, \ldots, P_n)$ is *O-perspective* with $(Q_1, \ldots, Q_n)$.

Dually, let $k \in \mathscr{L}$ and $P, Q \in \mathscr{P}$ such that $P, Q \oslash k$. Define λ by $\lambda x = (x \wedge k) \vee Q$ for $x \circ P$. We shall call λ a *perspectivity between P and Q with axis k* and write $\lambda: P \overset{k}{\overline{\wedge}} Q$. For λ a perspectivity between P and Q with axis k and for $l_i \circ P$ and $m_i = \lambda l_i$ for $i = 1, \ldots, n$, we shall write $(l_1, \ldots, l_n) \overset{k}{\overline{\wedge}} (m_1, \ldots, m_n)$ and say $(l_1, \ldots, l_n)$ is *k-perspective* with $(m_1, \ldots, m_n)$.

We note that if π is a perspectivity between l and m, then π is a one-to-one correspondence between the set of points on l and the set of points on m. Dually, if λ is a perspectivity between P and Q, then λ is a one-to-one correspondence between the set of lines on P and the set of lines on Q.

Let n be a positive integer. Let $O_i \in \mathscr{P}$, $l_0, l_i \in \mathscr{L}$ such that $O_i \oslash l_{i-1}$, l_i and $\pi_i : l_{i-1} \overset{O_i}{\overline{\wedge}} l_i$ for $i = 1, \ldots, n$. We shall call $\pi = \pi_n \pi_{n-1} \cdots \pi_2 \pi_1$ (that is, the composite of π_n through π_1) a *projectivity between l_0 and l_n* and write

$$\pi: l_0 \overset{O_1}{\overline{\wedge}} l_1 \overset{O_2}{\overline{\wedge}} l_2 \cdots l_{n-1} \overset{O_n}{\overline{\wedge}} l_n .$$

Dually, let $k_i \in \mathscr{L}$, $P_0, P_i \in \mathscr{P}$ such that $k_i \oslash P_{i-1}, P_i$ and $\lambda_i : P_{i-1} \overset{k_i}{\overline{\wedge}} P_i$ for $i = 1, \ldots, n$. We shall call $\lambda = \lambda_n \lambda_{n-1} \cdots \lambda_2 \lambda_1$ a *projectivity between P_0 and P_n* and write

$$\lambda: P_0 \overset{k_1}{\overline{\wedge}} P_1 \overset{k_2}{\overline{\wedge}} P_2 \cdots P_{n-1} \overset{k_n}{\overline{\wedge}} P_n .$$

We shall call π a *self-projectivity of l_0* if $l_0 = l_n$ and λ a *self-projectivity of P_0* if $P_0 = P_n$. A point $P \circ l_0$ (line $l \circ P_0$) will be said to be *fixed* if $\pi P = P$ ($\lambda l = 1$).

THEOREM 7.1. Let A, B, C be distinct points on a line l and A', B', C' distinct points on a line l'. Then there exists a projectivity π between l and l' such that $A' = \pi A$, $B' = \pi B$, and $C' = \pi C$. Moreover, π can be realized as a composite of as few as two or three perspectivities according as $l \neq l'$ or $l = l'$.

PROOF. Let A, B, C be distinct and on l, and A', B', C' distinct and on l'. Consider, first, the case when $l \neq l'$. Without loss of generality, let $A' \varnothing l$. Now let $l'' \circ A'$ such that $l'' \neq l', A \vee A'$ and $O \circ A \vee A'$ such that $O \neq A, A'$ (see Figure 7.1). Then $O \varnothing l, l''$. Define $B'' = (B \vee O) \wedge l''$, $C'' = (C \vee O) \wedge l''$, and $O' = (B'' \vee B') \wedge (C'' \vee C')$. Then $O' \varnothing l'', l'$. Further, $(A, B, C) \overset{O}{\doublebarwedge} (A', B'', C'') \overset{O'}{\doublebarwedge} (A', B', C')$, so that π given by $\pi\colon l \overset{O}{\doublebarwedge} l'' \overset{O'}{\doublebarwedge} l'$ will suffice. The case when $l = l'$ is left as an exercise.

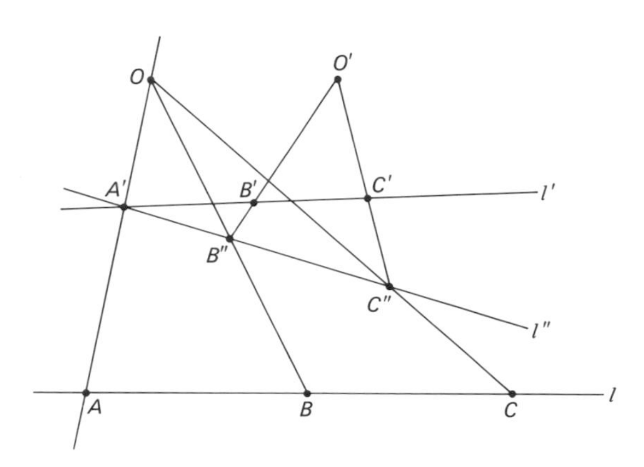

Figure 7.1

LEMMA 7.2. Let $(\mathscr{P}, \mathscr{L}, \circ)$ be desarguesian. If l, m, n are distinct concurrent lines, O, P points such that $O \varnothing l, m$ and $P \varnothing m, n$, and $\pi\colon l \overset{O}{\doublebarwedge} m \overset{P}{\doublebarwedge} n$, then π is a perspectivity between l and n.

PROOF. Let l, m, n be distinct and concurrent, $O \varnothing l, m$, and $P \varnothing m, n$. If $O = P, \pi\colon l \overset{O}{\doublebarwedge} n$. Let $O \neq P$. Then let $A \circ l$ such that $A \neq l \wedge m$ and $A \varnothing O \vee P$, and define $B = (A \vee O) \wedge m$, $C = (B \vee P) \wedge n$, and $Q = (A \vee C) \wedge (O \vee P)$ (see Figure 7.2). Now let $X \circ l$ such that $X \neq A$, $l \wedge m$. Define $Y = (X \vee O) \wedge m$ and $Z = (Y \vee P) \wedge n$. Then $\pi X = Z$. Applying the Theorem of Desargues, we see that $Z \circ X \vee Q$, so that $\pi\colon l \overset{Q}{\doublebarwedge} n$. Thus π is a perspectivity between l and n.

LEMMA 7.3. In a desarguesian projective plane, a projectivity between distinct lines given by a composite of three perspectivities can be realized as a composite of two perspectivities.

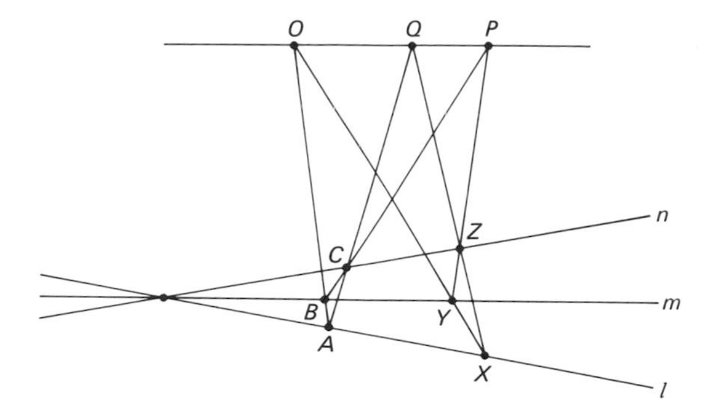

Figure 7.2

PROOF. Let $\pi : l \overset{O}{\doublebarwedge} m \overset{P}{\doublebarwedge} n \overset{Q}{\doublebarwedge} k$ with $l \neq k$. Consider, first, the case when there are at least five points on a line. We shall consider four cases as determined by conditions on O, P, Q, l, m, n, and k.

CASE 1. Let l, m, n, k be distinct, no three concurrent and $P \varnothing (l \wedge m) \vee (n \wedge k)$. Define $j = (l \wedge m) \vee (n \wedge k)$. Then $P \varnothing j$ and $\pi : l \overset{O}{\doublebarwedge} m \overset{P}{\doublebarwedge} j \overset{P}{\doublebarwedge} n \overset{Q}{\doublebarwedge} k$. [Note that the projectivity between m and n given by $m \overset{P}{\doublebarwedge} n$ equals that given by $m \overset{P}{\doublebarwedge} j \overset{P}{\doublebarwedge} n$ (see Figure 7.3)]. Now we apply Lemma 7.2 to the projectivity given by $l \overset{O}{\doublebarwedge} m \overset{P}{\doublebarwedge} j$. Since l, m, j are concurrent,

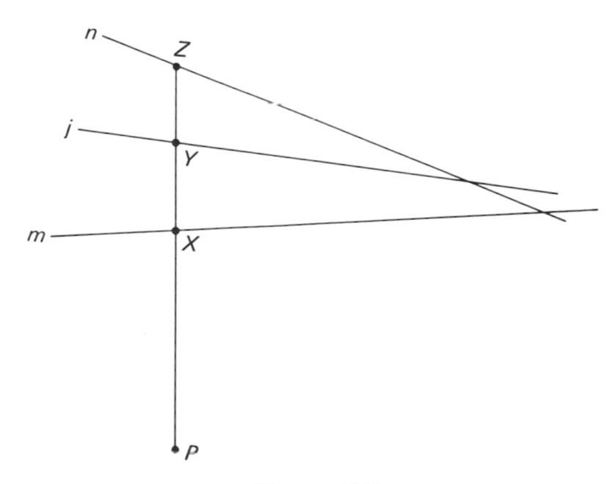

Figure 7.3

there exists O' such that this projectivity is given by $l \overset{O'}{\doublebarwedge} j$. Hence $\pi\colon l \overset{O'}{\doublebarwedge} j \overset{P}{\doublebarwedge} n \overset{Q}{\doublebarwedge} k$. Now since j, n, k are concurrent, we have, by a similar application of Lemma 7.2, $\pi\colon l \overset{O'}{\doublebarwedge} j \overset{Q'}{\doublebarwedge} k$ for some point Q'.

CASE 2. Let l, m, n, k be distinct, no three concurrent and $P \circ (l \wedge m) \vee (n \wedge k)$. Let $m' \in \mathscr{L}$ such that $m' \circ m \wedge n$, $m' \varnothing O$ and $m' \neq m, n$ (see Figure 7.4). Then $\pi\colon l \overset{O}{\doublebarwedge} m' \overset{O}{\doublebarwedge} m \overset{P}{\doublebarwedge} n \overset{Q}{\doublebarwedge} k$, so that $\pi\colon l \overset{O}{\doublebarwedge} m' \overset{P'}{\doublebarwedge} n \overset{Q}{\doublebarwedge} k$, by Lemma 7.2, for some point P'. Define $R = l \wedge m$, $R' = l \wedge m'$ and

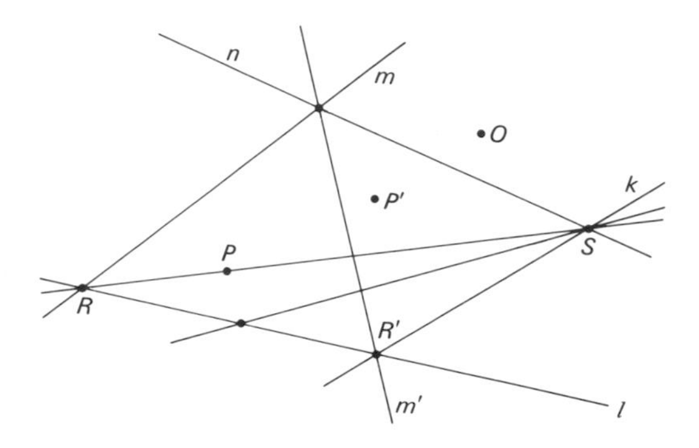

Figure 7.4

$S = n \wedge k$. Then $R \neq R'$ since $m \neq m'$, and $\pi R = S$ since $(R) \overset{O}{\doublebarwedge} (R) \overset{P}{\doublebarwedge} (S) \overset{Q}{\doublebarwedge} (S)$. Thus $\pi R' \neq S$ since π is one-to-one. Now $P' \varnothing R' \vee S$. Otherwise, if $P' \circ R' \vee S$, we should have $(R') \overset{O}{\doublebarwedge} (R') \overset{P'}{\doublebarwedge} (S) \overset{Q}{\doublebarwedge} (S)$, so that $\pi R' = S$, a contradiction. Hence we have π given by a composite of perspectivities satisfying the hypothesis of Case 1 provided that no three of l, m', n, and k are concurrent. We can realize this last condition if m' is chosen so that $m' \varnothing l \wedge k$.

CASE 3. Let l, m, n, k be distinct and at least three concurrent. Reduction of π to a composite of two perspectivities is immediate if all four lines or l, m, n or m, n, k are concurrent. Let l, n, k be concurrent. Let $n' \in \mathscr{L}$ such that $n' \circ m \wedge n$, $n' \varnothing Q$ and $n' \neq m, n$ (see Figure 7.5). Then $\pi\colon l \overset{O}{\doublebarwedge} m \overset{P}{\doublebarwedge} n \overset{Q}{\doublebarwedge} n' \overset{Q}{\doublebarwedge} k$. From this, we have, by Lemma 7.2, $\pi\colon l \overset{O}{\doublebarwedge} m \overset{P'}{\doublebarwedge}$

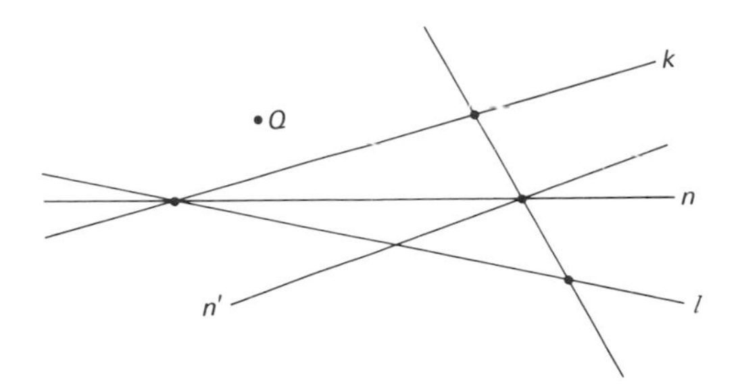

Figure 7.5

$n' \overset{Q}{\overline{\overline{\wedge}}} k$ for some point P' since m, n, n' are concurrent. Now no three of l, m, n', k are concurrent and we have reduced π to a composite of perspectivities satisfying the hypothesis of Case 1 or Case 2. The subcase for l, m, k concurrent would be treated similarly.

CASE 4. Let at least two of l, m, n, k be equal. Reduction of π to a composite of two perspectivities is immediate if three of the lines are equal or if $l = m$ or $m = n$ or $n = k$. Let $l = n$. Also let $n' \in \mathscr{L}$ such that $n' \circ m \wedge n, n' \varnothing Q$ and $n' \neq m, n$ (see Figure 7.6). Then $\pi : l \overset{O}{\overline{\overline{\wedge}}} m \overset{P}{\overline{\overline{\wedge}}} n \overset{Q}{\overline{\overline{\wedge}}} n' \overset{Q}{\overline{\overline{\wedge}}} k$, so that $\pi : l \overset{O}{\overline{\overline{\wedge}}} m \overset{P'}{\overline{\overline{\wedge}}} n' \overset{Q}{\overline{\overline{\wedge}}} k$ for some point P'. We have $l \neq n'$. If $n' \neq k$, we have expressed π as a composite of perspectivities satisfying the hypothesis of Case 3. Otherwise, if $n' = k$, the further reduction of π to a composite of two perspectivities is immediate. The subcase for $m = k$ would be treated similarly.

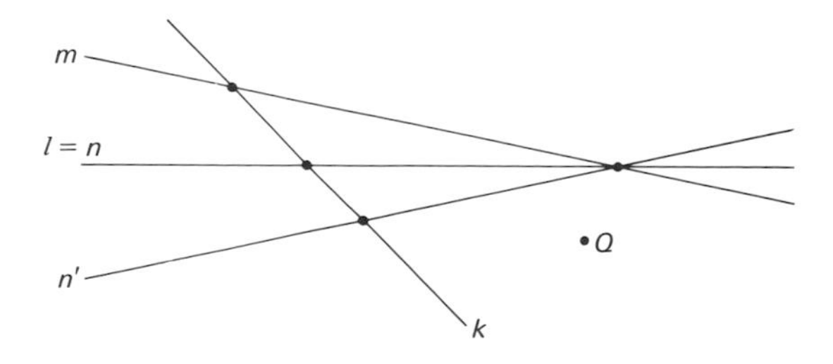

Figure 7.6

The remaining case, when there are less than five points on a line (that is, exactly three or four), is left as an exercise. This completes the proof.

THEOREM 7.4. In a desarguesian projective plane, a projectivity between lines l and m can be realized as a composite of no more than two or three perspectivities according as $l \neq m$ or $l = m$.

PROOF. We shall consider the case when $l \neq m$. (The remaining case is left as an exercise.) We shall proceed inductively. The result is immediate if we are given a projectivity which is a single perspectivity (or even the composite of two or three perspectivities). Let the result hold for $n (n \geqq 3)$, that is, a projectivity given as the composite of n perspectivities can be expressed as a composite of two. Now let π be the composite of $n+1$ perspectivities, say

$$\pi : l \overset{O_1}{\overline{\overline{\wedge}}} l_1 \overset{O_2}{\overline{\overline{\wedge}}} l_2 \overset{O_3}{\overline{\overline{\wedge}}} l_3 \overset{O_4}{\overline{\overline{\wedge}}} l_4 \cdots l_n \overset{O_{n+1}}{\overline{\overline{\wedge}}} m.$$

First, consider when $l \neq l_3$. Then, by Lemma 7.3,

$$\pi : l \overset{O_2'}{\overline{\overline{\wedge}}} l_2' \overset{O_3'}{\overline{\overline{\wedge}}} l_3 \overset{O_4}{\overline{\overline{\wedge}}} l_4 \cdots l_n \overset{O_{n+1}}{\overline{\overline{\wedge}}} m$$

for some points O_2', O_3' and line l_2'. Now, by the induction hypothesis, π can be expressed as the composite of two perspectivities. For the remaining case, let $l = l_3$. Now let $l_3' \in \mathscr{L}$ such that $l_3' \neq l, l_1$ and $l_3' \not\circ O_4$. Then

$$\pi : l \overset{O_1}{\overline{\overline{\wedge}}} l_1 \overset{O_2}{\overline{\overline{\wedge}}} l_2 \overset{O_3}{\overline{\overline{\wedge}}} l_3 \overset{O_4}{\overline{\overline{\wedge}}} l_3' \overset{O_4}{\overline{\overline{\wedge}}} l_4 \cdots l_n \overset{O_{n+1}}{\overline{\overline{\wedge}}} m.$$

By Lemma 7.3,

$$\pi : l \overset{O_1}{\overline{\overline{\wedge}}} l_1 \overset{O_2'}{\overline{\overline{\wedge}}} l_2' \overset{O_3'}{\overline{\overline{\wedge}}} l_3' \overset{O_4}{\overline{\overline{\wedge}}} l_4 \cdots l_n \overset{O_{n+1}}{\overline{\overline{\wedge}}} m$$

for some points O_2', O_3' and line l_2'. This case has been reduced to the previous case and we can consider the proof completed.

THEOREM 7.5. Let $(\mathscr{P}, \mathscr{L}, \circ)$ be desarguesian, Z, U, I distinct points on some line l and $(\mathscr{F}, +, \cdot)$ the division ring associated with (Z, U, I). Then each transformation π_i defined below is a self-projectivity of l.

(a) Let $A \in \mathscr{F}$. For $X \circ l$,

$$\pi_1 X = \begin{cases} X + A & \text{if } X \neq I, \\ I & \text{if } X = I. \end{cases}$$

(b) For $X \circ l$,

$$\pi_2 X = \begin{cases} -X & \text{if } X \neq I, \\ I & \text{if } X = I. \end{cases}$$

(c) Let $A \in \mathscr{F}$ such that $A \neq Z$. For $X \circ l$,

$$\pi_3 X = \begin{cases} A \cdot X & \text{if } X \neq I, \\ I & \text{if } X = I. \end{cases}$$

(d) Let $A \in \mathscr{F}$ such that $A \neq Z$. For $X \circ l$,

$$\pi_4 X = \begin{cases} X \cdot A & \text{if } X \neq I, \\ I & \text{if } X = I. \end{cases}$$

(e) For $X \circ l$,

$$\pi_5 X = \begin{cases} X^{-1} & \text{if } X \neq Z, I, \\ I & \text{if } X = Z, \\ Z & \text{if } X = I. \end{cases}$$

The proof is left as an exercise.

EXERCISES

7.1. Show that a perspectivity between lines l and m is a one-to-one correspondence between the set of points on l and the set of points on m. What can be said about π^{-1}?

7.2. If $\pi: l \overset{O}{\overline{\wedge}} l$ for $O \not\circ l$, then π is the identity transformation on the set of points on l.

7.3. Show that a composite of projectivities is a projectivity. What can be said about the inverse of a projectivity?

7.4. Complete the proof of Theorem 7.1. Also prove the dual of Theorem 7.1 directly, that is, without applying the Principle of Duality.

7.5. Where, in the proof of Lemma 7.3, did we use at least four points on a line? Five points? Complete the proof for the cases when there are exactly three or four points on a line.

7.6. Complete the proof of Theorem 7.4.

7.7. Prove Theorem 7.5. Determine the number of fixed points for each π_i. Describe π_i^{-1} for each i.

7.2. SOME CLASSICAL THEOREMS

We shall study three statements called "theorems." However, as with the case of the Theorem of Pappus, they will not be proved as theorems in a general projective plane. In fact, we shall show that each is equivalent to the Theorem of Pappus. Thus we shall have alternate characterizations of pappian projective planes.

FUNDAMENTAL THEOREM (OF PROJECTIVE GEOMETRY). Let distinct points A, B, C be on a line l and distinct points A', B', C' be on a line l'. Then there exists one and only one projectivity π between l and l' such that $A' = \pi A$, $B' = \pi B$ and $C' = \pi C$.

(THREE) FIXED POINT THEOREM. If a self-projectivity π of a line l has three fixed points, then π is the identity transformation on l.

CHARACTERIZATION THEOREM (OF PERSPECTIVITIES AMONG PROJECTIVITIES). Let π be a projectivity between l and m with $l \neq m$. Then π is a perspectivity if and only if $l \wedge m$ is fixed under π.

THEOREM 7.6. The following statements are equivalent:

(a) Theorem of Pappus,
(b) Fixed Point Theorem,
(c) Fundamental Theorem,
(d) Characterization Theorem.

PROOF The proof will consist of a cycle of four implications from which we can conclude the equivalence of the four statements.

PART 1. Theorem of Pappus implies Fixed Point Theorem. Let $(\mathscr{P}, \mathscr{L}, \circ)$ be pappian. In addition, let π be a self-projectivity of a line l such that F, G, H are distinct fixed points on l. We know that $(\mathscr{P}, \mathscr{L}, \circ)$ is desarguesian, so that we can apply Theorem 7.4 to π to obtain $\pi\colon l \overset{O'}{\doublebarwedge} m'' \overset{P''}{\doublebarwedge} n' \overset{Q}{\doublebarwedge} l$ (see Figure 7.7). Without loss of generality, let $H \not\circ m'', n'$. Before proceeding, we shall show that the perspectivities describing π

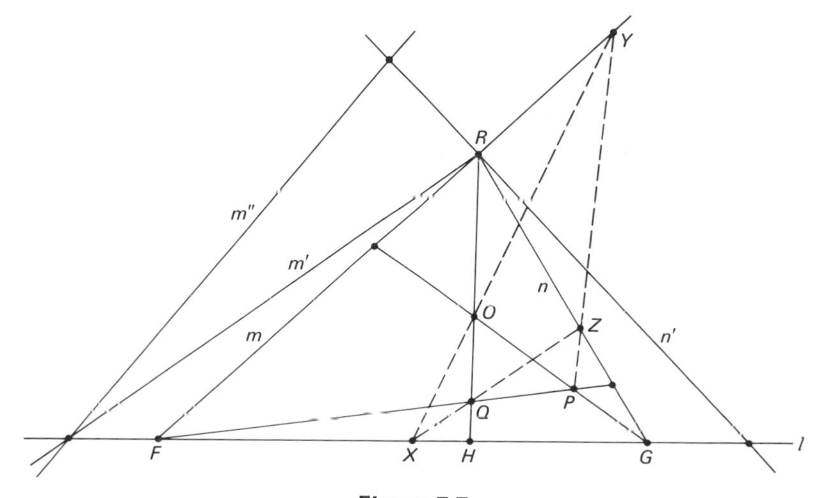

Figure 7.7

can be selected in a special way. If $m'' \wedge n' \circ Q \vee H$, let $m' = m''$. Otherwise, define $R = (Q \vee H) \wedge n'$ and $m' = R \vee (m'' \wedge l)$. If $P'' \circ m'$, we have $\pi(m'' \wedge l) = H$, that is, $m'' \wedge l = H$ contrary to $H \not\circ m''$. Thus $P'' \not\circ m'$. Now $\pi\colon l \overset{O'}{\doublebarwedge} m'' \overset{P''}{\doublebarwedge} m' \overset{P''}{\doublebarwedge} n' \overset{Q}{\doublebarwedge} l$, so that $\pi\colon l \overset{O}{\doublebarwedge} m' \overset{P''}{\doublebarwedge} n' \overset{Q}{\doublebarwedge} l$ for some point O. Further, $O \circ Q \vee H$ since $\pi H = H$. If $F \circ m'$, let $m = m'$. Otherwise, define $m = F \vee R$. Then $O \not\circ m$ and $\pi\colon l \overset{O}{\doublebarwedge} m \overset{O}{\doublebarwedge} m' \overset{P''}{\doublebarwedge} n' \overset{Q}{\doublebarwedge} l$, so that $\pi\colon l \overset{O}{\doublebarwedge} m \overset{P'}{\doublebarwedge} n' \overset{Q}{\doublebarwedge} l$ for some point P'. If $G \circ n'$, let $n = n'$. Otherwise, define $n = G \vee R$. Then $Q \not\circ n$ and $\pi\colon l \overset{O}{\doublebarwedge} m \overset{P'}{\doublebarwedge} n' \overset{Q}{\doublebarwedge} n \overset{Q}{\doublebarwedge} l$, so that $\pi\colon l \overset{O}{\doublebarwedge} m \overset{P}{\doublebarwedge} n \overset{Q}{\doublebarwedge} l$ for some point P. In addition, $P \circ F \vee Q$ since $\pi F = F$, and $P \circ G \vee O$ since $\pi G = G$. We have the form we want for π.

Now let $X \circ l$ such that $X \neq F, G, H$. Define $Y = (X \vee O) \wedge m$ and $Z = (Y \vee P) \wedge n$. Consider the pappian point-cycle $\overline{PYORGF}$. Then, by the Theorem of Pappus,

$$(P \vee Y) \wedge (R \vee G) = Z,$$
$$(Y \vee O) \wedge (G \vee F) = X,$$
$$(O \vee R) \wedge (F \vee P) = Q$$

are collinear. Hence $(X) \overset{O}{\barwedge} (Y) \overset{P}{\barwedge} (Z) \overset{Q}{\barwedge} (X)$. Thus $\pi X = X$.

PART 2 Fixed Point Theorem implies Fundamental Theorem. Let $(\mathscr{P}, \mathscr{L}, \circ)$ satisfy the Three Fixed Point Theorem. Let A, B, C be distinct points on l and A', B', C' be distinct points on l'. By Theorem 7.1, there exists a projectivity ρ between l and l' such that $A' = \rho A$, $B' = \rho B$, and $C' = \rho C$. Let σ be a projectivity between l and l' such that $A' = \sigma A$, $B' = \sigma B$, and $C' = \sigma C$. Consider the composite $\pi = \rho^{-1}\sigma$ which is a self-projectivity of l with fixed points A, B, C. Thus, by the Fixed Point Theorem, π is the identity on the set of points on l. Now for $X \circ l$,

$$\rho X = \rho(\pi X) = \rho(\rho^{-1}(\sigma X)) = \sigma X.$$

Thus $\rho = \sigma$ and therefore ρ is unique.

PART 3. Fundamental Theorem implies Characterization Theorem. Let $(\mathscr{P}, \mathscr{L}, \circ)$ satisfy the Fundamental Theorem. Let π be a projectivity between distinct lines l and m. If π is a perspectivity, it is immediate that $1 \wedge m$ is fixed. For the converse, let $l \wedge m$ be a fixed point of π. Let $P, Q \circ l$ such that $P, Q, l \wedge m$ are distinct (see Figure 7.8). Define $O = (P \vee \pi P) \wedge (Q \vee \pi Q)$. Then the perspectivity ρ between l and m with center O satisfies $\rho P = \pi P$, $\rho Q = \pi Q$ and $\rho(l \wedge m) = l \wedge m = \pi(l \wedge m)$. Thus, by the Fundamental Theorem, $\rho = \pi$, so that π is a perspectivity.

PART 4. Characterization Theorem implies Theorem of Pappus. Let $(\mathscr{P}, \mathscr{L}, \circ)$ satisfy the Characterization Theorem. Let $\overline{PQ'RP'QR'}$ be a pappian point-cycle (Figure 7.9). Define $R'' = (P \vee Q') \wedge (P' \vee Q)$, $Q'' = (P' \vee R) \wedge (P \vee R')$, $l'' = R'' \vee Q''$, $P'' = (Q \vee R') \wedge l''$, $A = (P \vee Q) \wedge l''$, $B = (P' \vee Q) \wedge (P \vee R')$ and $C = (P \vee Q') \wedge (P' \vee R)$. Let $\pi\colon (P \vee Q') \overset{P'}{\barwedge} (P \vee R') \overset{Q}{\barwedge} l''$. Then $(P, C, Q', R'') \overset{P'}{\barwedge} (P, Q'', R', B) \overset{Q}{\barwedge}$

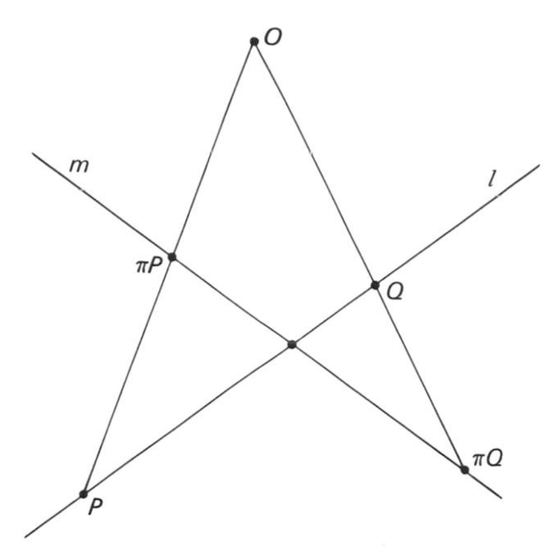

Figure 7.8

(A, Q'', P'', R''). Hence $\pi R'' = R''$. Now since $R'' = (P \vee Q') \wedge l''$, we know that π is a perspectivity by the Characterization Theorem. The center of π is $(P \vee \pi P) \wedge (C \vee \pi C) = (P \vee A) \wedge (C \vee Q'') = R$. Thus $R \circ (Q' \vee \pi Q') = (Q' \vee P'')$. Hence $(Q' \vee R) \wedge (Q \vee R') = P'' \circ l''$. This completes the proof of the theorem.

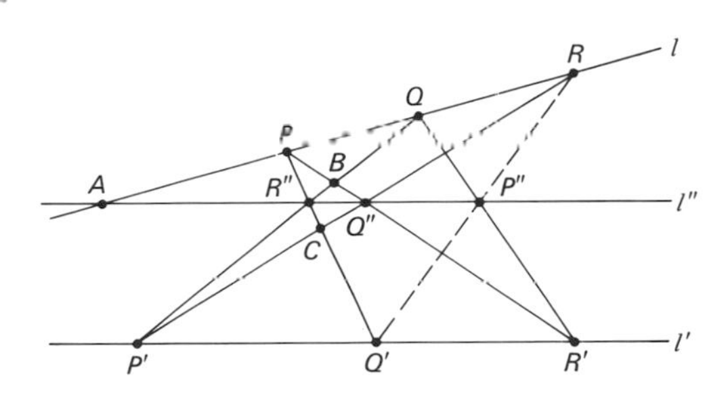

Figure 7.9

COROLLARY 7.7. In a pappian projective plane, a self-projectivity has either none, one, two, or all points fixed.

The proof is left as an exercise.

EXERCISES

7.8. Prove Corollary 7.7.
7.9. Find examples of self-projectivities showing that each case in the corollary can occur.

7.3. A NONPAPPIAN EXAMPLE

In Chapter 5, we saw that a pappian projective plane is desarguesian and suggested that the converse does not necessarily hold. In Chapter 6, we indicated that a finite desarguesian projective plane is pappian. Hence, if an example of a desarguesian projective plane exists which is not pappian, necessarily it is not finite. In particular, if the example is an algebraic incidence basis with elements from a division ring where multiplication is noncommutative, the division ring is not finite.

We shall construct such an example using an infinite division ring with a noncommutative multiplication. This will be the division ring of quaternions. (See Appendix C for a brief discussion of quaternions.)

THEOREM 7.8. There exists a desarguesian projective plane which is not pappian.

PROOF. Let $(\mathscr{D}, +, \cdot)$ be the division ring of quaternions and $(\mathscr{P}_1, \mathscr{L}_1, \circ_1)$ the algebraic incidence basis with elements from $(\mathscr{D}, +, \cdot)$. Since $(\mathscr{D}, +, \cdot)$ is a division ring, we have, by Theorem 6.13, that $(\mathscr{P}_1, \mathscr{L}_1, \circ_1)$ is a desarguesian projective plane. To show that it is nonpappian, we shall show that the Fixed Point Theorem fails to hold. Then the desired conclusion follows from Theorem 7.6. Let $l = [1, -1, 0]$. Then the points on l are of the form $(1, 1, 0)$ or $(\mathbf{x}, \mathbf{x}, 1)$ for $\mathbf{x} \in \mathscr{D}$. Define π by $\pi(1, 1, 0) = (0, 0, 1)$, $\pi(0, 0, 1) = (1, 1, 0)$, and for $\mathbf{x} \neq 0$, $\pi(\mathbf{x}, \mathbf{x}, 1) = (-\mathbf{x}^{-1}, -\mathbf{x}^{-1}, 1)$. Then π is a transformation with domain consisting of the set of points on l and range contained in this set. Let $Z = (0, 0, 1)$, $U = (1, 1, 1)$, $J = (1, 0, 0)$, and $K = (0, 1, 0)$. (Note that $l = Z \vee U$.) Now consider the addition and multiplication of points on l associated with the coordinate system (Z, U, J, K). We shall show that for $A = (\mathbf{a}, \mathbf{a}, 1)$ with $\mathbf{a} \in \mathcal{D}$, $-A \equiv (-\mathbf{a}, -\mathbf{a}, 1)$ and $A^{-1} \equiv (\mathbf{a}^{-1}, \mathbf{a}^{-1}, 1)$ provided $\mathbf{a} \neq 0$. To this end, note that we have the following auxiliary elements $j = [0, 1, 0]$, $k = [1, 0, 0]$, $l = [1, -1, 0]$, $m = [0, 0, 1]$, and $I = (1, 1, 0)$. Let $A = (\mathbf{a}, \mathbf{a}, 1)$ for $\mathbf{a} \in \mathcal{D}$. In case $\mathbf{a} = 0$, we have $-A = -Z = Z \equiv (0, 0, 1) = (-0, -0, 1)$. Let $\mathbf{a} \neq 0$ (see Figure 7.10). Now we shall list several statements where each can be justified by simple algebraic computations to show the incidence of a

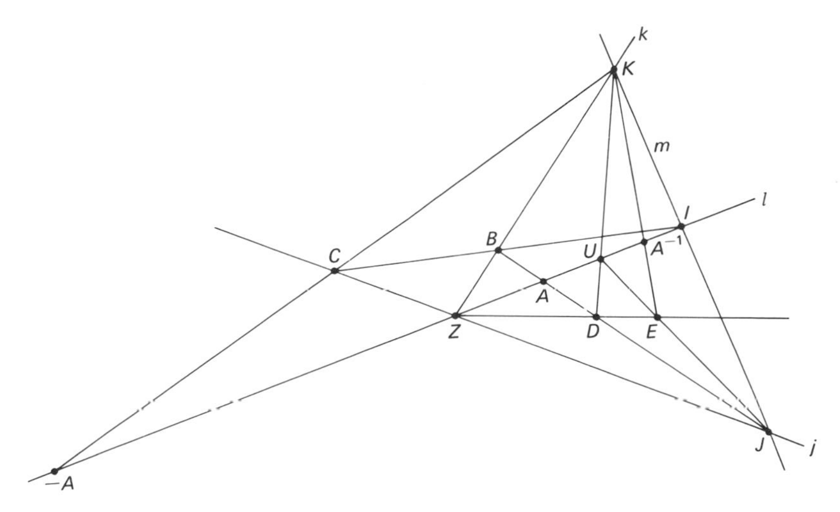

Figure 7.10

line and two previously determined points, or dually, the incidence of a point and two previously determined lines. We have $J \vee A \equiv [0, -1, \mathbf{a}]$, $B = (J \vee A) \wedge k' \equiv (0, \mathbf{a}, 1)$, $I \vee B \equiv [1, -1, \mathbf{a}]$, $C = (I \vee B) \wedge j' \equiv (\mathbf{a}, 0, -1)$, $K \vee C \equiv [1, 0, \mathbf{a}]$, and $-A = (K \vee C) \wedge l \equiv (-\mathbf{a}, -\mathbf{a}, 1)$. Now in case $\mathbf{a} = 1$, we have $A^{-1} = U^{-1} = U \equiv (1, 1, 1) = (1^{-1}, 1^{-1}, 1)$. Let $\mathbf{a} \neq 1$. That $A^{-1} \equiv (\mathbf{a}^{-1}, \mathbf{a}^{-1}, 1)$ is left as an exercise. Now we are able to conclude the following about π:

$$\pi A = \begin{cases} -A^{-1} & \text{for } A \neq Z, I, \\ I & \text{for } A = Z, \\ Z & \text{for } A = I. \end{cases}$$

The justification of this and the fact that π is a self-projectivity is left as an exercise. Now $\pi(\mathbf{i}, \mathbf{i}, 1) = (-\mathbf{i}^{-1}, -\mathbf{i}^{-1}, 1) = (\mathbf{i}, \mathbf{i}, 1)$, so that $(\mathbf{i}, \mathbf{i}, 1)$ is a fixed point. Similarly, we have $(\mathbf{j}, \mathbf{j}, 1)$ and $(\mathbf{k}, \mathbf{k}, 1)$ fixed. However, $\pi(1, 1, 1) = (-1^{-1}, -1^{-1}, 1) = (-1, -1, 1) \not\equiv (1, 1, 1)$. Hence $(1, 1, 1)$ is not fixed. Thus the Fixed Point Theorem fails to hold, and therefore $(\mathscr{P}_1, \mathscr{L}_1, \circ_1)$ is nonpappian.

EXERCISE

7.10. Complete the proof of Theorem 7.8.

HARMONIC QUADRUPLES

CHAPTER 8

When looking at a euclidean plane for the motivation of new concepts to be studied in a projective plane, we have not considered directly the notions of "length" or "distance." We shall consider them presently to introduce the ideas studied in this chapter, which will give us, then, a broader foundation for the study of the real projective plane in Chapter 9.

Consider, in a euclidean plane, a line segment AB and points C, D that divide AB in the same ratio, that is, $\overline{AC}/\overline{CB} = -\overline{AD}/\overline{DB}$. (The symbol $\overline{XY}$ is the directed length of segment XY relative to some number scale selected on the line on XY. See Figure 8.1.) This relationship on four collinear points is invariant under projections. If π is a projection between lines l and l' and $A' = \pi A$, $B' = \pi B$, $C' = \pi C$, and $D' = \pi D$, then $\overline{A'C'}/\overline{C'B'} = -\overline{A'D'}/\overline{D'B'}$ (where the number scale on l' can be selected arbitrarily). Hence this relationship on four points is eligible to be considered under our scheme of considering those notions invariant under projections. However, we must first equivalently express this relationship using euclidean notions that have an analogy in our projective theory.

First, let us note that the condition $\overline{AC}/\overline{BC} = -\overline{AD}/\overline{DB}$ holds if and only if the real numbers $\overline{AC}$, $\overline{AB}$, $\overline{AD}$ are in harmonic progression.

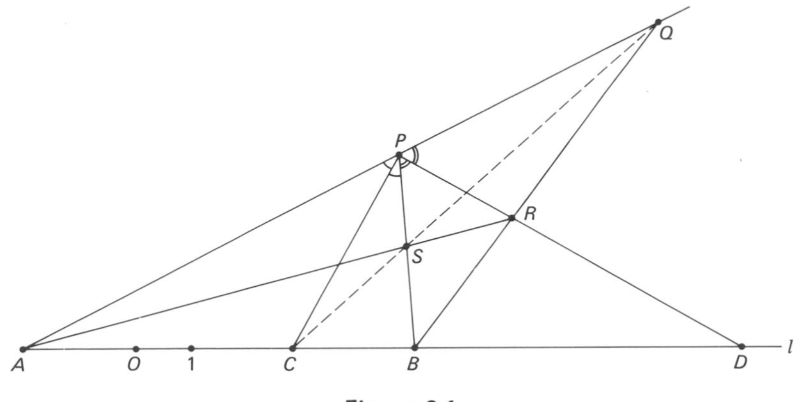

Figure 8.1

Secondly, if PC is the interior bisector of angle APB in triangle APB, then PD is an exterior bisector if and only if C, D divide the segment AB in the same ratio (see Figure 8.1). We still are employing euclidean notions for which we have no projective analogy.

Now let us consider triangle APB with points C, D on side AB such that PC is the interior bisector of angle APB and PD is an exterior bisector. Let R be on the line PD, Q be the intersection of lines AP and BR, and S be the intersection of lines BP and AR. Then it can be shown that C, S, Q are collinear (and conversely). Now we have A, B as diagonal points of complete quadrangle $PQRS$, C, D on the remaining two sides of $PQRS$ and A, B, C, D collinear. Thus we, now, have the relationship on A, B, C, D equivalently expressed in terms of euclidean notions that do have an analogy in our projective theory. This is the basis for the relation on collinear quadruples of points that we shall study.

Throughout the remainder of this chapter $(\mathscr{P}, \mathscr{L}, \circ)$ is to be a projective plane.

8.1. FANO AXIOM

As indicated in the prefatory remarks, we are interested in the join l of two diagonal points of a complete quadrangle and the points of intersection of l and the remaining two sides of the complete quadrangle. In particular, we want these remaining sides to intersect l in distinct points (that is, we are interested in the study of four specific points on l). However, this does not occur in general—a fanian plane is an example to the contrary. Thus we shall introduce, as needed, an additional condition (sometimes called the *Fano Axiom*) on the projective planes we are going to study.

AXIOM F. The diagonal points of a complete quadrangle are not collinear.

THEOREM 8.1. There exists an example of a pappian projective plane satisfying Axiom F.

The proof is left as an exercise. Note that we have chosen to include the pappian condition in the consistency statement since we shall be working, ultimately, with Axiom F in pappian projective planes. The independence of Axiom F from the axioms for a pappian projective plane is immediate if we consider a fanian plane as an independence incidence basis.

THEOREM 8.2. A set of axioms for a projective plane together with Axiom F is self-dual.

The proof is left as an exercise.

EXERCISES

8.1. Prove Theorem 8.1. (*Hint:* See Exercises 4.16 and 5.19.)
8.2. Prove Theorem 8.2.
8.3. What would hold for addition relative to a coordinate system (Z, U, J, K) if the diagonal points of every complete quadrangle were collinear? (*Hint*: Consider $U + U$.)

8.2. HARMONIC QUADRUPLES

Throughout this section, we shall assume that $(\mathscr{P}, \mathscr{L}, \circ)$ is desarguesian and satisfies Axiom F.

We define a relation $\mathbf{H} \subset \mathscr{P} \times \mathscr{P} \times \mathscr{P} \times \mathscr{P}$ by the following. Let A, B, C, D be collinear points. Then $(A, B, C, D) \in \mathbf{H}$ if A, B are diagonal points of some complete quadrangle and C, D are on the remaining two sides of the complete quadrangle. If $(A, B, C, D) \in \mathbf{H}$, we shall write $\mathbf{H}(AB, CD)$ and say that the ordered quadruple (A, B, C, D) is *harmonic* (see Figure 8.2).

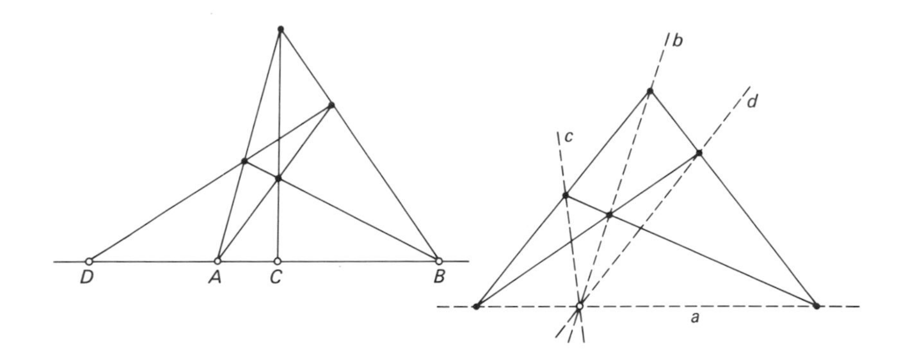

Figure 8.2

Dually, $\mathbf{h} \subset \mathscr{L} \times \mathscr{L} \times \mathscr{L} \times \mathscr{L}$ is defined by the following. Let a, b, c, d be concurrent lines. Then $(a, b, c, d) \in \mathbf{h}$ if a, b are diagonal lines of some complete quadrilateral and c, d are on the remaining two vertices. If $(a, b, c, d) \in \mathbf{h}$, we shall write $\mathbf{h}(ab, cd)$ and say that (a, b, c, d) is *harmonic* (see Figure 8.2).

COROLLARY 8.3. Let $A, B, C, D \in \mathscr{P}$ such that $\mathbf{H}(AB, CD)$. Then

(a) A, B, C, D are distinct;

(b) $\mathbf{H}(AB, DC)$, $\mathbf{H}(BA, CD)$, and $\mathbf{H}(BA, DC)$.

The proof is left as an exercise.

THEOREM 8.4. If A, B, C are distinct collinear points, then there exists one and only one point D such that $\mathbf{H}(AB, CD)$.

The proof is left as an exercise.

THEOREM 8.5. If $A, B, C, D \in \mathscr{P}$ such that $\mathbf{H}(AB, CD)$, then $\mathbf{H}(CD, AB)$.

PROOF. Let $\mathbf{H}(AB, CD)$ and let A, B be diagonal points of complete quadrangle $PQRS$ with C, D on the remaining two sides. Now let T be the remaining diagonal point (see Figure 8.3). Define $U = (C \vee Q) \wedge$

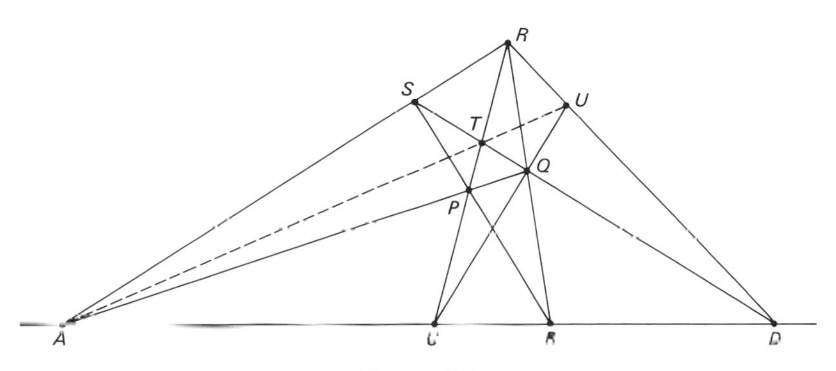

Figure 8.3

$(D \vee R)$. Then triangles PST and QRU are perspective from $A \vee B$. Hence by the dual of the Theorem of Desargues, $S \vee R$, $P \vee Q$, and $T \vee U$ are concurrent; in particular, $A \circ T \vee U$. Now C, D are diagonal points of complete quadrangle $QURT$ and A, B are on the remaining two sides. Thus $\mathbf{H}(CD, AB)$.

COROLLARY 8.6. If $A, B, C, D \in \mathscr{P}$ such that $\mathbf{H}(AB, CD)$, then $\mathbf{H}(AB, DC)$, $\mathbf{H}(BA, CD)$, $\mathbf{H}(BA, DC)$, $\mathbf{H}(CD, AB)$, $\mathbf{H}(CD, BA)$, $\mathbf{H}(DC, AB)$, and $\mathbf{H}(DC, BA)$.

The question naturally arises whether some permutation of $\{A, B, C, D\}$ other than those appearing in the corollary, say (A, C, B, D), can be harmonic when $\mathbf{H}(AB, CD)$. This question cannot be answered in general (see Exercise 8.7). In the next chapter, we shall consider projective planes so restrictive that only the permutations of $\{A, B, C, D\}$ listed in the corollary can be harmonic when $\mathbf{H}(AB, CD)$.

THEOREM 8.7. Let A, B, C, D be distinct collinear points and a, b, c, d be distinct concurrent lines such that $A \circ a$, $B \circ b$, $C \circ c$, and $D \circ d$. Then $\mathbf{H}(AB, CD)$ if and only if $\mathbf{h}(ab, cd)$.

PROOF. Let $A, B, C, D \circ l$ and $a, b, c, d \circ P$ such that $A \circ a$, $B \circ b$, $C \circ c$, and $D \circ d$. For the "only if" part, let $\mathbf{H}(AB, CD)$ (see Figure 8.4). Let $Q \circ b$ such that $Q \neq P, B$. Define $R = (A \vee Q) \wedge c$. Then

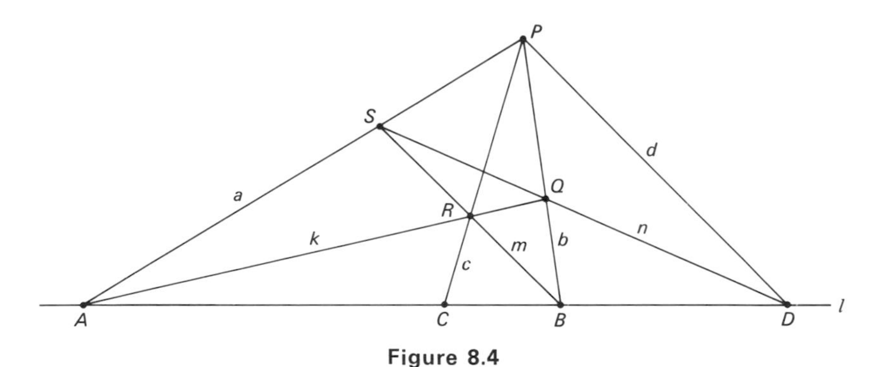

Figure 8.4

$R \neq A, Q$. Define $S = a \wedge (B \vee R)$. Then $\{P, Q, R, S\}$ is a quadrangle. Moreover, A, B are diagonal points of complete quadrangle $PQRS$ and C is on one of the remaining sides, so that by Theorem 8.4, $D \circ Q \vee S$. Define $k = A \vee Q$, $m = S \vee B$, and $n = S \vee D$. Then $\{k, l, m, n\}$ is a quadrilateral. Moreover, a, b are diagonal sides and c, d are on the remaining two vertices. Hence $\mathbf{h}(ab, cd)$. The converse follows dually.

LEMMA 8.8. Let A, B, C, D be distinct points on l, $A', B', C', D' \circ l'$ and $O \not\circ l, l'$ such that $(A, B, C, D) \overset{O}{\overline{\wedge}} (A', B', C', D')$. If $\mathbf{H}(AB, CD)$, then $\mathbf{H}(A'B', C'D')$.

The proof is left as an exercise.

THEOREM 8.9. Let π be a projectivity between lines l and m and let $\mathbf{H}(AB, CD)$ for distinct points A, B, C, D on l. Then $\mathbf{H}((\pi A)(\pi B), (\pi C)(\pi D))$.

The proof is left as an exercise.

We are able to derive a result in our present theory which is the direct analogy of one discussed in the prefatory remarks. Let Z, U, I be distinct points on some line l and $(\mathscr{F}, +, \cdot)$ the division ring associated with (Z, U, I). For $A, B, C \in \mathscr{F}$, we shall say that

(a) (A, B, C) is an *arithmetic triple* if $B - A = C - B$;
(b) (A, B, C) is an *harmonic triple* if (A^{-1}, B^{-1}, C^{-1}) is an arithmetic triple provided $A, B, C \neq Z$.

THEOREM 8.10. Let Z, U, I be distinct collinear points, $(\mathscr{F}, +, \cdot)$ be the division ring associated with (Z, U, I) and B, C, D be distinct elements in $\mathscr{F}$ different from Z. Then $\mathbf{H}(ZB, CD)$ if and only if (C, B, D) is an harmonic triple.

PROOF. Let Z, U, I be distinct collinear points and $(\mathscr{F}, +, \cdot)$ the division ring associated with (Z, U, I). Now let B, C, D be distinct elements in $\mathscr{F}$ different from Z. We shall consider, first, the case when $B = U$. For the "only if" part, let $\mathbf{H}(ZU, CD)$. Then $\mathbf{H}(DC, UZ)$. Now let $J, K \in \mathscr{P}$ such that (Z, U, J, K) is a coordinate system with $(\mathscr{F}, +, \cdot)$ as its associated division ring (see Figure 8.5). Successively apply Theorems 7.5 and 8.9 to show that the following statements are equivalent:

$$\mathbf{H}(DC, UZ); \quad \mathbf{H}(U(D^{-1} \cdot C), D^{-1}Z); \quad \mathbf{H}(C^{-1}D^{-1}, (D^{-1} \cdot C^{-1})Z);$$
$$\mathbf{H}(CD, (C \cdot D)I).$$

Now consider the complete quadrangle $JKLM$ where $L : (C, D)$ and $M : (D, C)$. Define $P = (L \vee M) \wedge l$. We have $\mathbf{H}(CD, PI)$ and therefore $P = C \cdot D$ by Theorem 8.4. Hence $C \cdot D \circ L \vee M$. But $L \vee M$ has

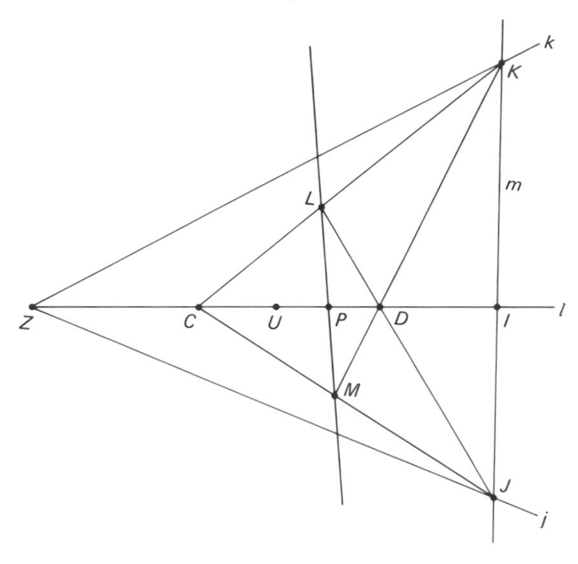

Figure 8.5

equation $Y = (-U) \cdot X + (C + D)$, so that $C \cdot D = (-U) \cdot (C \cdot D) + (C + D)$. Thus $C \cdot D + C \cdot D = C + D$. Now

$$\begin{aligned} C^{-1} + D^{-1} &= C^{-1} \cdot (U + C \cdot D^{-1}) \\ &= C^{-1} \cdot ((D + C) \cdot D^{-1}) \\ &= C^{-1} \cdot ((C \cdot D + C \cdot D) \cdot D^{-1}) \\ &= C^{-1} \cdot (C + C) \\ &= U + U. \end{aligned}$$

Hence $U - C^{-1} = D^{-1} - U$. Thus (C^{-1}, U, D^{-1}) is an arithmetic triple and (C, U, D) is an harmonic triple. The converse part of this case and the general case when B is not necessarily U are left as an exercise.

THEOREM 8.11. Let $(\mathscr{P}, \mathscr{L}, \circ)$ be pappian. Let A, B, X, Y be distinct points on a line l and π a self-projectivity of l such that $\pi A = A$, $\pi B = B$, and $\pi X = Y$. Then $\mathbf{H}(AB, XY)$ if and only if $\pi Y = X$.

PROOF. Let π be a self-projectivity of l such that $\pi A = A$, $\pi B = B$, $\pi X = Y$, where A, B, X, Y are distinct. For the "only if" part, let $\mathbf{H}(AB, XY)$. Let $PQRS$ be a complete quadrangle with A, B diagonal points and X, Y on the remaining two sides (see Figure 8.6). Define

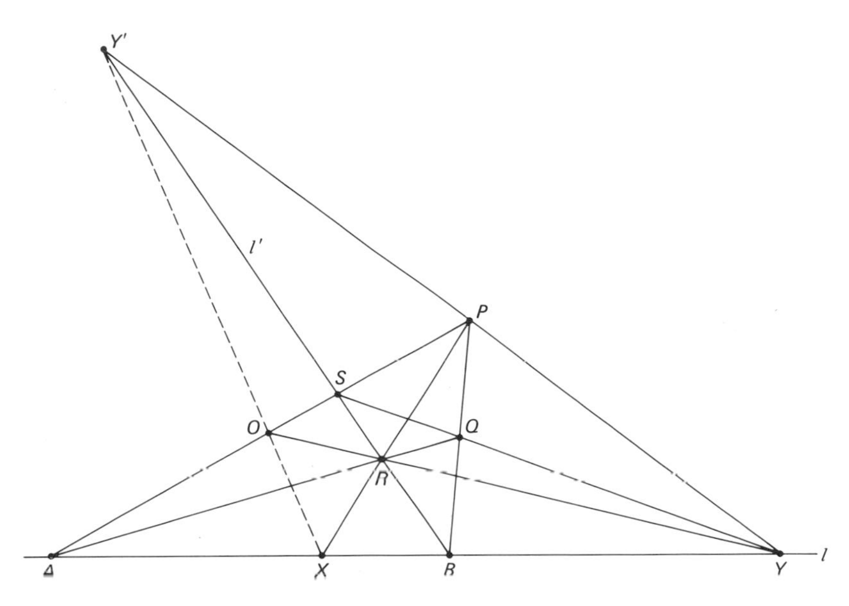

Figure 8.6

$l' = R \vee S$, $O = (Y \vee R) \wedge (P \vee S)$, and $Y' = (Y \vee P) \wedge l'$. Then triangles PRY and BAS are perspective from Q, whence from a line, in particular, the line $X \vee Y'$. Thus $O \circ X \vee Y'$. Now $(A, B, X, Y) \overset{P}{\doublebarwedge} (S, B, R, Y') \overset{O}{\doublebarwedge} (A, B, Y, X)$. Hence, by the Fundamental Theorem, $\pi\colon l \overset{P}{\doublebarwedge} l' \overset{Q}{\doublebarwedge} l$. Thus $\pi Y = X$.

For the converse, let $\pi Y = X$. Define $Y' = l' \wedge (Y \vee P)$, $S = l' \wedge (A \vee P)$, $R = l' \wedge (X \vee P)$, $Q = (A \vee R) \wedge (B \vee P)$, and $O = (A \vee P) \wedge (R \vee Y)$. Now $(A, B, X) \overset{P}{\doublebarwedge} (S, B, R) \overset{O}{\doublebarwedge} (A, B, Y)$. Thus by the

Fundamental Theorem, $\pi : l \overset{P}{\overline{\overline{\wedge}}} l' \overset{O}{\overline{\overline{\wedge}}} l$. In addition, $(Y) \overset{P}{\overline{\overline{\wedge}}} (Y') \overset{O}{\overline{\overline{\wedge}}} (\pi Y)$. But $\pi Y = X$. Hence, Y', O, X are collinear. Now triangles PRY and BAS are perspective from the line $O \vee X$, whence from a point, in particular, the point $(P \vee B) \wedge (R \vee A) = Q$. Hence $Y \circ Q \vee S$. Now we have A, B as diagonal points of complete quadrangle $PQRS$ and X, Y on the remaining two sides, so that $\mathbf{H}(AB, XY)$.

EXERCISES

8.4. Prove Corollary 8.3.

8.5. Prove Theorem 8.4.

8.6. Draw configurations illustrating the duals of Theorems 8.4 and 8.5.

8.7. Study the possibility or impossibility that $\mathbf{H}(AB, CD)$ and $\mathbf{H}(AC, BD)$ for distinct collinear points A, B, C, D in the projective plane obtained by imbedding each of the following:

(a) Basis 2.6;

(b) the geometry of the paper considered as an affine plane.

8.8. Draw a configuration illustrating the converse part of theorem 8.7.

8.9. Prove Lemma 8.8.

8.10. Prove Theorem 8.9.

8.11. Complete the proof of Theorem 8.10.

8.12. In Figure 8.7, consider the geometry of the paper as an affine plane

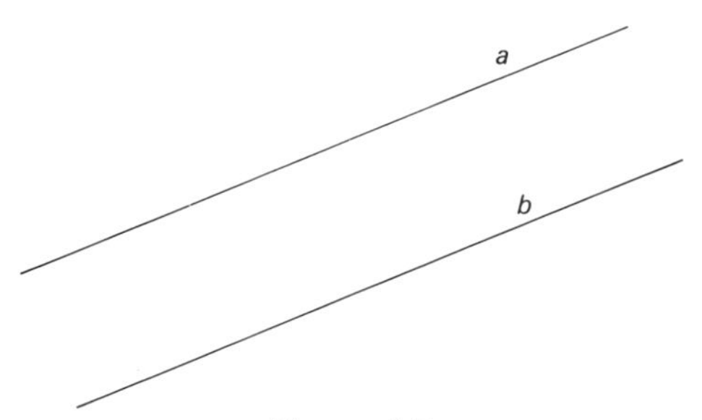

Figure 8.7

imbedded into a desarguesian projective plane. If $a \parallel b$, draw c so that $\mathbf{h}(\{a\}\{b\}, \{c\}\{l_\infty\})$.

8.13. Apply Theorem 8.11 to the applicable projectivities of Theorem 7.5. What are your conclusions?

THE REAL PROJECTIVE PLANE CHAPTER 9

We should like, at this time, to relate our study to that of euclidean planes. In the introductory chapter, we spoke briefly of studying euclidean geometry using primitive concepts and axioms not necessarily those used by Hilbert, in particular and insofar as it is possible, using primitive concepts enjoying a certain "invariance" under projections. We began our study with affine planes. Then we continued in the environment of projective planes after describing fully the relationship between these two concepts. A euclidean plane was presented as an example of an affine plane (Basis 2.1) and a deletion subgeometry of a projective plane was shown to be an affine plane. It would seem natural, then, to try to determine those projective planes whose deletion subgeometries are euclidean planes or, stated another way, to characterize the projective plane obtained by imbedding a euclidean plane. Of course, we should want the characterizing properties to involve only concepts with euclidean analogs which enjoy the desired "invariance" under projections. Our characterization will depend on the relationship, established in the study of analytic geometry, that exists between a euclidean plane and the system of real numbers.

As noted in an earlier comment, the analytic study of a euclidean plane deals, essentially, with an isomorphism between Bases 2.1 and 2.7.

(In addition, other euclidean concepts, not all necessarily "invariant" under projections, are given an analytic description based on a special choice of a cartesian coordinate system.) We also saw, earlier, that Basis 3.5, the algebraic incidence basis with elements from the field of real numbers, has a deletion subgeometry isomorphic to Basis 2.7. Further, we saw that any pappian projective plane can be coordinatized using a field constructed from the elements of the projective plane. It will be our project, now, to specialize the pappian projective planes considered so that the field obtained will be the field of real numbers. The fields obtained thus far lack the algebraic concept of "ordering."

Hilbert realized an ordering in his euclidean studies by considering a "betweenness" relation for triples of collinear points. In euclidean space, betweenness is "preserved" under a projection between two planes when the planes are parallel, but not necessarily when they are intersecting. To illustrate this latter possibility, consider intersecting planes $\mathscr{A}$ and $\mathscr{A}'$ and point O not on $\mathscr{A}$ and $\mathscr{A}'$. Let A, B, C, D be on a line l on $\mathscr{A}$ such that B is between A and C, C is between B and D, and the line on O and C is parallel to $\mathscr{A}'$. Further, let A', B', D', and l' be the correspondents of A, B, D, and l, respectively, under the projection from $\mathscr{A}$ to $\mathscr{A}'$ through O (Figure 9.1). We see that C has no correspondent on $\mathscr{A}'$. Hence we cannot speak of the preservation of betweenness for the correspondents of any triple of points on $\mathscr{A}$ that includes C. However, if we consider the projective plane obtained by imbedding the euclidean plane determined by O and l, we can think of C as having a correspondent C', namely, the ideal point associated with l'. It would appear, now, that the associate of B' is between C' and the associate of A'; but this would also appear to be the case for the associate of D' since the addition of C', in a sense, "closes" l' in a "continuous loop." (We have encountered this "endless" or "circular" quality of projective lines before; for example, in Basis 3.2, a projective line was a great circle on a sphere. Exercises 6.5 and 6.13 also gave an indication of this quality.) Let us view, instead, the pair $\{A', C'\}$ as "separating" the pair $\{B', D'\}$. This interpretation can be applied also to pairs $\{A, C\}$ and $\{B, D\}$ since the ideal point associated with l also "closes" l in a "loop." It is this "separating" relation (with ideal points included) that we consider "invariant" under our projection.

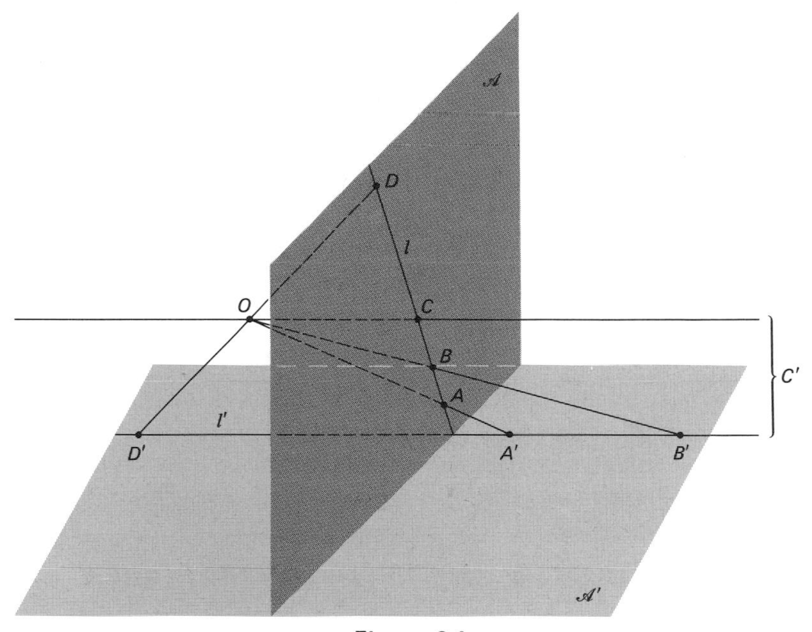

Figure 9.1

The above discussion suggests how we might proceed in our projective development. We shall include among our primitive concepts a relation pairing pairs of points or, equivalently, a relation consisting of ordered quadruples of points.

9.1. SEPARATION

The system $(\mathscr{P}, \mathscr{L}, \circ, \mathbf{S})$ will be called an *incidence basis with a separation relation* if $(\mathscr{P}, \mathscr{L}, \circ)$ is an incidence basis and $\mathbf{S}$ is a nonempty subset of $\mathscr{P} \times \mathscr{P} \times \mathscr{P} \times \mathscr{P}$. An incidence basis with a separation relation $(\mathscr{P}, \mathscr{L}, \circ, \mathbf{S})$ will be called a *separated projective plane* if $(\mathscr{P}, \mathscr{L}, \circ)$ is a projective plane and the axioms stated below are satisfied. Before the statement of the axioms, however, we shall introduce a notational convention similar to that introduced for the harmonic relation $\mathbf{H}$. We shall write $\mathbf{S}(AB, CD)$ if $(A, B, C, D) \in \mathbf{S}$.

AXIOM S_1. If $S(AB, CD)$, then A, B, C, D are distinct and collinear.

AXIOM S_2. If A, B, C, D are distinct collinear points, then $S(AB, CD)$ or $S(AC, DB)$ or $S(AD, BC)$.

AXIOM S_3. If $S(AB, CD)$, then $S(AB, DC)$.

AXIOM S_4. If $S(AB, CD)$ and $S(AC, BE)$, then $S(AC, DE)$.

AXIOM S_5. If $S(AB, CD)$ and A', B', C', D' are collinear points such that $(A, B, C, D) \overset{O}{\overline{\overline{\wedge}}} (A', B', C', D')$ for some point O, then $S(A'B', C'D')$.

Throughout the remainder of this section, we shall assume that $(\mathscr{P}, \mathscr{L}, \circ, S)$ is a separated pappian projective plane. Then we have the following immediate consequence of Axiom S_5.

COROLLARY 9.1. If $S(AB, CD)$ and π is a projectivity between $A \vee B$ and a line l, then $S((\pi A)(\pi B), (\pi C)(\pi D))$.

THEOREM 9.2. If $S(AB, CD)$, then $S(AB, DC)$, $S(BA, CD)$, $S(BA, DC)$, $S(CD, AB)$, $S(CD, BA)$, $S(DC, AB)$, and $S(DC, BA)$.

PROOF. Let $S(AB, CD)$. First, we shall show that $S(CD, AB)$. To this end, let $P \not\circ A \vee B$ and $Q \circ P \vee A$ such that $Q \neq P, A$ (see Figure 9.2). Define $R = (P \vee C) \wedge (Q \vee D)$, $S = (P \vee B) \wedge (Q \vee D)$, and $T = (P \vee B) \wedge (A \vee R)$. Then

$$(A, B, C, D) \overset{P}{\overline{\overline{\wedge}}} (Q, S, R, D) \overset{A}{\overline{\overline{\wedge}}} (P, S, T, B) \overset{R}{\overline{\overline{\wedge}}} (C, D, A, B).$$

Thus by Corollary 9.1, $S(CD, AB)$. The remaining possibilities now follow from this result and Axiom S_3.

THEOREM 9.3. If A, B, C, D are distinct collinear points, then exactly one of the statements $S(AB, CD)$, $S(AC, DB)$, $S(AD, BC)$ holds.

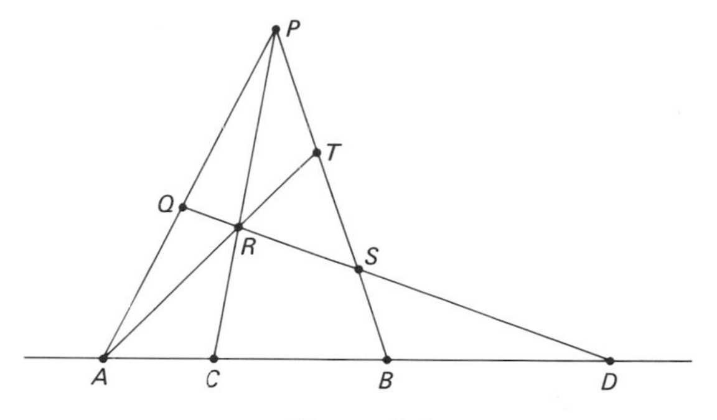

Figure 9.2

PROOF. We have at least one of the statements holding by Axiom S_2. Suppose that $S(AB, CD)$ and $S(AC, DB)$. Then by Theorem 9.2, $S(AB, CD)$ and $S(AC, BD)$, so that by Axiom S_4, $S(AC, DD)$ contrary to Axiom S_1. Similarly, it can be shown that the remaining two combinations of statements are contradictory. Thus exactly one of the statements holds.

THEOREM 9.4. If A, B, C are distinct collinear points, then there exists a point D such that $S(AB, CD)$.

PROOF. Let A, B, C be distinct points on a line l. Let $l' \in \mathscr{L}$ and let A', B', C', D' be distinct points on l'. (There exist at least four points on some line since $S \neq \varnothing$, so that there exist at least four points on each line.) By Axiom S_2, one of $S(A'B', C'D')$, $S(A'C', D'B')$, $S(A'D', B'C')$ holds. Without loss of generality, let $S(A'B', C'D')$ hold. Define π to be the projectivity between l' and l such that $\pi A' = A$, $\pi B' = B$, and $\pi C' = C$. Define $D = \pi D'$. Then by Corollary 9.1, $S(AB, CD)$.

LEMMA 9.5. If A, B, C, D, E are distinct collinear points, then all of the statements $S(AB, CD)$, $S(AB, DE)$, and $S(AB, EC)$ cannot hold.

PROOF. Let A, B, C, D, E be distinct and collinear. Suppose $\mathbf{S}(AB, CD)$, $\mathbf{S}(AB, DE)$, and $\mathbf{S}(AB, EC)$. Now one of $\mathbf{S}(AC, DE)$, $\mathbf{S}(AD, EC)$, or $\mathbf{S}(AE, CD)$ holds. Let $\mathbf{S}(AC, DE)$. Then we have $\mathbf{S}(CD, AB)$ and $\mathbf{S}(CA, DE)$. Hence by Axiom S_4, $\mathbf{S}(CA, BE)$ contrary to Theorem 9.3 since $\mathbf{S}(AB, EC)$. In a similar way, each of the hypotheses $\mathbf{S}(AD, EC)$ and $\mathbf{S}(AE, CD)$ leads to a contradiction. The details are left as an exercise. Thus not all of $\mathbf{S}(AB, CD)$, $\mathbf{S}(AB, DE)$ and $\mathbf{S}(AB, EC)$ can hold.

THEOREM 9.6. Axiom F is satisfied. Moreover, if A, B, C, D are distinct collinear points such that $\mathbf{H}(AB, CD)$, then $\mathbf{S}(AB, CD)$.

PROOF. Let $PQRS$ be a complete quadrangle. Define $A = (P \vee Q) \wedge (R \vee S)$, $B = (P \vee S) \wedge (Q \vee R)$, $l = A \vee B$, $C = (P \vee R) \wedge l$, and $D = (Q \vee S) \wedge l$ (Figure 9.3). By Theorem 9.4 there exists $T \circ R \vee S$

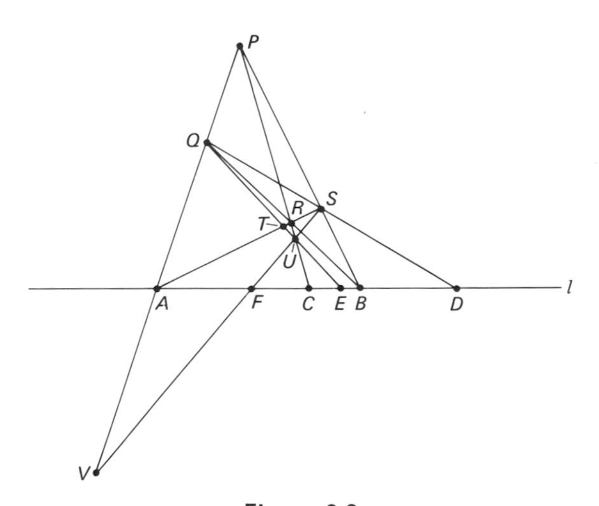

Figure 9.3

such that $\mathbf{S}(AR, ST)$. Define $U = (Q \vee T) \wedge (P \vee R)$, $V = (P \vee Q) \wedge (S \vee U)$, $E = (Q \vee T) \wedge l$, and $F = (S \vee U) \wedge l$. Then $(A, R, S, T) \overset{Q}{\overline{\overline{\wedge}}} (A, B, D, E)$, and therefore by Axiom S_5, $\mathbf{S}(AB, DE)$. In addition,

$(A, R, S, T) \overset{U}{\bar{\bar{\wedge}}} (A, P, V, Q) \overset{S}{\bar{\bar{\wedge}}} (A, B, F, D)$, so that by Corollary 11.1, $\mathbf{S}(AB, FD)$. Finally, $(A, R, S, T) \overset{U}{\bar{\bar{\wedge}}} (A, C, F, E)$. Hence $\mathbf{S}(AC, FE)$. Now by Lemma 9.5, $\mathbf{S}(AB, EF)$ cannot hold since we have $\mathbf{S}(AB, DE)$ and $\mathbf{S}(AB, FD)$. Thus by Axiom S_2, $\mathbf{S}(AF, BE)$ or $\mathbf{S}(AE, FB)$ holds. Let $\mathbf{S}(AF, BE)$ hold. Then $\mathbf{S}(FE, AC)$ and $\mathbf{S}(FA, EB)$. Hence by Axiom S_4, $\mathbf{S}(FA, CB)$. Now $\mathbf{S}(AF, CB)$ and $\mathbf{S}(AB, FD)$ so that by Axiom S_4 again, $\mathbf{S}(AB, CD)$. The argument for the alternate case, when $\mathbf{S}(AE, FB)$, follows similarly since E and F enter the discussion symmetrically. Now by Axiom S_1, $C \neq D$. Hence $(P \vee R) \wedge (Q \vee S) \varnothing l$. Thus Axiom F is satisfied. The proof of the second part of the theorem is left as an exercise.

THEOREM 9.7. If $\mathbf{S}(AB, CE)$ and $\mathbf{S}(AC, DE)$, then $\mathbf{S}(AB, DE)$.

The proof is left as an exercise.

We close this section with a brief discussion of duality. The basis $(\mathscr{P}, \mathscr{L}, \circ, \mathbf{S})$ is not self-dual since $\mathbf{S}$ is not self-dual and its dual concept is not present. However, this lack of duality is readily remedied by the introduction of a concept dual to $\mathbf{S}$, a nonempty relation $\mathbf{s} \subset \mathscr{L} \times \mathscr{L} \times \mathscr{L} \times \mathscr{L}$. Define

$$\mathbf{s} = \{(j, k, m, n) : j, k, m, n \text{ are lines concurrent on some point } P \text{ such that } \mathbf{S}((j \wedge l)(k \wedge l), (m \wedge l)(n \wedge l)) \text{ for some line } l \varnothing P\}.$$

First, we must show that $\mathbf{s}$ is well defined insofar as the definition is independent of the choice of the line l not on P. Now the dual statements of Axioms S_1–S_5 must be verified. The details are left as an exercise.

EXERCISES

9.1. Prove Theorem 9.7.

9.2. Complete the proof of Lemma 9.5.

9.3. Complete the proof of Theorem 9.6.

9.4. Complete the discussion of dualizing the separation.

9.2. ORDERED FIELDS

One procedure for introducing an order relation for a field is to consider, first, the set of "positive" elements. Such a set can be defined abstractly by properties like the properties stated in the next theorem. This is used, then, to define an order relation and to derive needed properties.

Throughout this section $(\mathscr{P}, \mathscr{L}, \circ, \mathbf{S})$ is to be a separated pappian projective plane. Let (Z, U, J, K) be a coordinate system and $(\mathscr{F}, +, \cdot)$ its associated field. Define

$$\mathscr{G} = \{x \in \mathscr{F} : X = U \quad \text{or} \quad \mathbf{S}(ZU, XI) \quad \text{or} \quad \mathbf{S}(ZX, UI)\}.$$

THEOREM 9.8.

(a) If $X \in \mathscr{F}$, then exactly one of $X \in \mathscr{G}$, $X = Z$ or $-X \in \mathscr{G}$ holds.
(b) If $X, Y \in \mathscr{G}$, then $X + Y \in \mathscr{G}$.
(c) If $X, Y \in \mathscr{G}$, then $X \cdot Y \in \mathscr{G}$.

PROOF. For part (a), let $X \in \mathscr{F}$ such that $X \neq Z$. In case $X = U$, $X \in \mathscr{G}$. Let $X \neq U$. Then one of $\mathbf{S}(ZU, XI)$, $\mathbf{S}(ZX, UI)$, $\mathbf{S}(ZI, UX)$ holds. If either of the first two possibilities holds, $X \in \mathscr{G}$. Let $\mathbf{S}(ZI, UX)$. If $X = -U$, $-X = U \in \mathscr{G}$. Let $X \neq -U$. Suppose $\mathbf{S}(ZI, X(-U))$. We have $\mathbf{S}(ZI, (-U)U)$ since $\mathbf{H}(ZI, (-U)U)$. This contradicts Lemma 9.5. Hence $\mathbf{S}(ZI, X(-U)$ cannot hold. Thus $\mathbf{S}(ZX, (-U)I)$ or $\mathbf{S}(Z(-U), XI)$. Then by Theorem 7.5(b) and Corollary 9.1, we have $\mathbf{S}(Z(-X), UI)$ or $\mathbf{S}(ZU, (-X)I)$, respectively. In both cases, $-X \in \mathscr{G}$. Thus we have, for $X \in \mathscr{F}$, at least one of $X \in \mathscr{G}$, $X = Z$ or $-X \in \mathscr{G}$ holding. To show that not more than one can hold, note that $X \neq Z$ if $X \in \mathscr{G}$ or $-X \in \mathscr{G}$. Suppose $X, -X \in \mathscr{G}$. There are nine possible ways this hypothesis can be realized. It is left as an exercise to show that each leads to a contradiction. Hence we cannot have both $X, -X \in \mathscr{G}$. This completes the proof of part (a) of the theorem

For part (c), let $X, Y \in \mathscr{G}$. If $X = U$ or $Y = U$, it is immediate that $X \cdot Y \in \mathscr{G}$. Let $X, Y \neq U$. There are three essential cases to be considered.

· CASE 1. Let $\mathbf{S}(ZU, XI)$ and $\mathbf{S}(ZU, YI)$. Then by Theorem 7.5(d) and Corollary 9.1, $\mathbf{S}(ZY, (X \cdot Y)I)$. Thus by Theorem 9.7, $\mathbf{S}(ZU, (X \cdot Y)I)$, so that $X \cdot Y \in \mathscr{G}$.

CASE 2. Let $\mathbf{S}(ZU, XI)$ and $\mathbf{S}(ZY, UI)$. Then by Theorem 7.5(e), Corollary 9.1, and Theorem 9.2, $\mathbf{S}(ZX^{-1}, UI)$ and $\mathbf{S}(ZU, Y^{-1}I)$. In case $X = Y^{-1}$, $X \cdot Y = Y^{-1} \cdot Y = U \in \mathscr{G}$. Let $X \neq Y^{-1}$. Now one of $\mathbf{S}(ZI, Y^{-1}X)$, $\mathbf{S}(ZY^{-1}, XI)$, $\mathbf{S}(ZX, Y^{-1}I)$ must hold. It cannot be $\mathbf{S}(ZI, Y^{-1}\mathscr{G})$; otherwise, $\mathbf{S}(ZU, IX)$ and $\mathbf{S}(ZI, Y^{-1}X)$ implies $\mathbf{S}(ZU, Y^{-1}X)$ contrary to Lemma 9.5. If $\mathbf{S}(ZY^{-1}, XI)$ holds, then by Theorem 7.5(d) and Corollary 9.1, $\mathbf{S}(ZU, (X \cdot Y)I)$, so that $X \cdot Y \in \mathscr{G}$. Similarly if $\mathbf{S}(ZX, Y^{-1}I)$ holds, then $\mathbf{S}(Z(X \cdot Y), UI)$, and therefore $X \cdot Y \in \mathscr{G}$.

CASE 3. Let $\mathbf{S}(ZX, UI)$ and $\mathbf{S}(ZY, UI)$. This case is left as an exercise. The proof of part (c) will be considered complete.

Before proceeding with the proof of part (b), let us consider the proof of the following statement:

(1) If $A, B \in \mathscr{G}$ such that $\mathbf{S}(ZB, AI)$, then

$$\mathbf{S}(Z(A + B), BI) \quad \text{and} \quad \mathbf{S}(Z(A + B), AI).$$

Let $\mathbf{S}(ZB, AI)$. Then $A \neq Z$. In addition, by Theorem 7.5(b) and Corollary 9.1, $\mathbf{S}(Z(-B), (-A)I)$. Thus by Theorem 7.5(a) and Corollary 9.1, $\mathbf{S}(BZ, (B - A)I)$, so that

$$\mathbf{S}(I(B - A), BZ). \tag{2}$$

We have $\mathbf{S}(ZI, A(-A))$ since $\mathbf{H}(ZI, A(-A))$. Thus

$$\mathbf{S}(BI, (B + A)(B - A)),$$

and therefore

$$\mathbf{S}(IB, (B - A)(B + A)). \tag{3}$$

Now by Axiom S_4, (3) and (2) imply

$$\mathbf{S}(I(B - A), (B + A)Z). \tag{4}$$

One of $\mathbf{S}(ZB, (B + A)I)$, $\mathbf{S}(ZI, B(B + A))$, $\mathbf{S}(Z(B + A), BI)$ must hold. Suppose $\mathbf{S}(ZB, (B + A)I)$. Then $\mathbf{S}(I(B + A), BZ)$ and from (3), $\mathbf{S}(IB, (B + A)(B - A))$. Thus by Axiom S_4, $\mathbf{S}(IB, Z(B - A))$ contrary to (2). Suppose $\mathbf{S}(ZI, B(B+A))$. Then $\mathbf{S}(IZ, B(B+A))$ and from (3), $\mathbf{S}(IB, (B-A)(B+A))$. Thus by Theorem 9.7, $\mathbf{S}(IZ, (B-A)(B+A))$ contrary to (4). Therefore we must have $\mathbf{S}(Z(B+A), BI)$, so that $\mathbf{S}(Z(A+B), BI)$. It is left as an exercise to show that $\mathbf{S}(Z(A+B), AI)$.

Finally, for the proof of part (b), let $X, Y \in \mathscr{G}$. We shall consider two cases.

CASE 1. Let $X = U$. For $Y = U$, we have $\mathbf{S}(Z(U+U), UI)$ since $\mathbf{H}(Z(U+U), UI)$. Thus $U + U \in \mathscr{G}$. If $Y \neq U$, we have $\mathbf{S}(ZU, YI)$ or $\mathbf{S}(ZY, UI)$. Hence by (1), $\mathbf{S}(Z(Y+U), UI)$ or $\mathbf{S}(Z(U+Y), UI)$, respectively. Thus $X + Y \in \mathscr{G}$ in either case.

CASE 2. Let $X \neq U$. Then $\mathbf{S}(ZU, XI)$ or $\mathbf{S}(ZX, UI)$. Hence by Theorem 7.5(e) and Corollary 9.1, we have $\mathbf{S}(IU, X^{-1}Z)$ or $\mathbf{S}(IX^{-1}, UZ)$, respectively, that is, $\mathbf{S}(ZX^{-1}, UI)$ or $\mathbf{S}(ZU, X^{-1}I)$. Thus $X^{-1} \in \mathscr{G}$. By part (c), $X^{-1} \cdot Y \in \mathscr{G}$. It is left as an exercise to show that $U + X^{-1} \cdot Y \in \mathscr{G}$. Now $X + Y = X \cdot (U + X^{-1} \cdot Y) \in \mathscr{G}$. This completes the proof of Theorem 9.8.

Define $<$, a relation on $\mathscr{F} \times \mathscr{F}$, by

$$< \; = \{(X, Y) : X, Y \in \mathscr{F} \quad \text{such that} \quad Y - X \in \mathscr{G}\}.$$

If $(X, Y) \in \; <$, we shall write $X < Y$ (read " X is less than Y").

COROLLARY 9.9. For $X \in \mathscr{F}$, $X \in \mathscr{G}$ if and only if $Z < X$.

THEOREM 9.10.

(a) If $X \in \mathscr{F}$, $X \not< X$.
(b) If $X, Y, W \in \mathscr{F}$ such that $X < Y$ and $Y < W$, then $X < W$.
(c) If $X, Y \in \mathscr{G}$, then exactly one of $X = Y$, $X < Y$, or $Y < X$ holds.
(d) If $X, Y, W \in \mathscr{F}$ such that $X < Y$, then $X + W < Y + W$.
(e) If $X, Y, W \in \mathscr{F}$ such that $X < Y$ and $Z < W$, then $X \cdot W < Y \cdot W$.

Parts (a), (b), and (c) of the theorem are used, respectively, to make the following definitions:

(a) $<$ is said to be *irreflexive*;
(b) $<$ is said to be *transitive*;
(c) $<$ is said to be *trichotomous*.

An irreflexive, transitive, and trichotomous relation $<$ on any cartesian product $\mathbf{F} \times \mathbf{F}$ is said to be an *order relation* in $\mathbf{F}$. A system $(\mathbf{F}, +, \cdot, <)$, where $(\mathbf{F}, +, \cdot)$ is a field and $<$ is an order relation in $\mathbf{F}$ satisfying (d) and (e) of the theorem is said to be an *ordered field*. The theorem states that $(\mathscr{F}, +, \cdot, <)$ is an ordered field.

PROOF OF THEOREM 9.10. For part (a), let $X \in \mathscr{F}$. Suppose $X < X$. Then $Z = X - X \in \mathscr{G}$ contrary to Theorem 9.8. For part (b), let X, Y, $W \in \mathscr{F}$ such that $X < Y$ and $Y < W$. Then $Y - X$, $W - Y \in \mathscr{G}$. Now $W - X = (W - Y) + (Y - X) \in \mathscr{G}$. Hence $X < W$. The proof of the remaining parts is left as an exercise.

We now state two theorems to be used primarily in the next section, although they bear, in themselves, some interest and shed some light on the relationship between our separation and order relations.

THEOREM 9.11. Let $A \in \mathscr{G}$. Then $A < B$ if and only if $\mathbf{S}(ZB, AI)$.

PROOF. Let $A \in \mathscr{G}$. We shall consider the converse part first. Let $\mathbf{S}(ZB, AI)$. By Theorem 7.5(a) and Corollary 9.1, $\mathbf{S}((-A)(B - A), ZI)$, so that $\mathbf{S}(ZI, (B - A)(-A))$. Since $A \in \mathscr{G}$, we have $A \neq Z$, $-A \neq Z$, and $-A \notin \mathscr{G}$. Hence $\mathbf{S}(ZI, (-A)U)$. Therefore by Lemma 9.5, $\mathbf{S}(ZI, U(B - A))$ cannot hold. Hence if $B - A \neq U$, $\mathbf{S}(ZU, (B - A)I)$, or $\mathbf{S}(Z(B - A), UI)$, so that $B - A \in \mathscr{G}$. Thus $A < B$.

For the "only if" part, let $A < B$. Since $A \in \mathscr{G}$, we have $Z < A$ and therefore by the transitive property $Z < B$. Hence $B \in \mathscr{G}$. Suppose $\mathbf{S}(ZA, BI)$. Then by the above argument, $B < A$ contrary to the trichotomous property. Hence $\mathbf{S}(ZA, BI)$ cannot hold. Suppose $\mathbf{S}(ZI, AB)$. Then $A \neq U$; otherwise, if $A = U$, $\mathbf{S}(ZI, UB)$ contrary to $B \in \mathscr{G}$. Since $A \in \mathscr{G}$, $\mathbf{S}(ZU, AI)$ or $\mathbf{S}(ZA, UI)$. Let $\mathbf{S}(ZU, AI)$. Then $\mathbf{S}(IA, ZU)$ and $\mathbf{S}(IZ, AB)$ implies $\mathbf{S}(IZ, UB)$ contrary to $B \in \mathscr{G}$. Let $\mathbf{S}(ZA, UI)$. Then $\mathbf{S}(ZA, IU)$ and $\mathbf{S}(ZI, AB)$ implies $\mathbf{S}(ZI, BU)$ contrary to $B \in \mathscr{G}$. Hence $\mathbf{S}(ZI, AB)$ cannot hold. Thus $\mathbf{S}(ZB, AI)$.

THEOREM 9.12. Let $A, B, C \in \mathscr{F}$ such that $A < C$. Then
(a) $Z < A < B < C$ if and only if $\mathbf{S}(ZB, AC)$ and $A \in \mathscr{G}$;
(b) $A < B < C$ if and only if $\mathbf{S}(AC, BI)$.

The proof is left as an exercise.

We make a notational comment at this point. We shall write $X \leqq Y$ if $X < Y$ or $X = Y$ for $X, Y \in \mathscr{F}$.

EXERCISES

9.5. Complete the proof of part (a) of Theorem 9.8.
9.6. Complete the proof of part (c) of Theorem 9.8.
9.7. Show that $\mathbf{S}(ZB, AI)$ implies $\mathbf{S}(Z(A + B), AI)$.
9.8. In the proof of part (b) of Theorem 9.8, show that

$$U + X^{-1} \cdot Y \in \mathscr{G}.$$

9.9. Prove parts (c)–(e) of Theorem 9.10.
9.10. Study a combination of Theorems 9.7 and 9.11.
9.11. Prove Theorem 9.12

9.3. COMPLETENESS AND THE REAL NUMBERS

Let $(\mathscr{P}, \mathscr{L}, \circ, \mathbf{S})$ be a separated pappian projective plane. Our objective, in this section, is to state an axiom of *completeness* (or *continuity*) which will guarantee that the ordered field associated with any coordinate system in $(\mathscr{P}, \mathscr{L}, \circ, \mathbf{S})$ will be the system of real numbers. We shall assume the characterization of the system of real numbers as a *complete ordered field*, the latter being described below (see Eves and Newsom [9] for this characterization).

Our approach to completeness in the geometry will follow the approach of Richard Dedekind for real numbers (see Eves and Newsom [9]). We recall the notation $P(l)$ for $l \in \mathscr{L} : P(l) = \{Q \in \mathscr{P} : Q \circ l\}$.

Let $l \in \mathscr{L}$. We shall say that $\{\mathscr{A}, \mathscr{A}'\}$ is a *section on* l if $\mathscr{A}, \mathscr{A}' \subset P(l)$ such that

(a) there exist at least two points in each of $\mathscr{A}$ and $\mathscr{A}'$;
(b) $P(l) = \mathscr{A} \cup \mathscr{A}'$;
(c) if $X, Y \in \mathscr{A}$, $X \neq Y$ and $X', Y' \in \mathscr{A}'$, $X' \neq Y'$, then $\mathbf{S}(XX', YY')$ or $\mathbf{S}(XY', X'Y)$.

It follows that $\mathscr{A} \cap \mathscr{A}' = \varnothing$ for $\{\mathscr{A}, \mathscr{A}'\}$ a section.

Let $\{\mathscr{A}, \mathscr{A}'\}$ be a section on l and let $A, A' \circ l$ such that $A \neq A'$. We shall say that the section $\{\mathscr{A}, \mathscr{A}'\}$ is *at* the pair $\{A, A'\}$ if $\mathbf{S}(AA', XX')$ whenever $X \in \mathscr{A}$, $X' \in \mathscr{A}'$ such that X, X', A, A' are distinct (see Figure 9.4). Such a pair $\{A, A'\}$ is unique as stated in the next theorem.

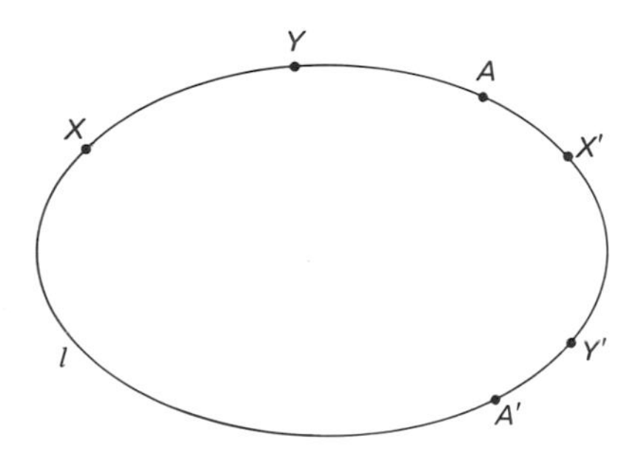

Figure 9.4

THEOREM 9.13. Let $A, A', B, B' \circ l$ such that $A \neq A'$ $B \neq B'$ and let $\{\mathscr{A}, \mathscr{A}'\}$ be a section on l at $\{A, A'\}$ and $\{B, B'\}$. Then $\{A, A'\} = \{B, B'\}$.

PROOF. Let $\{\mathscr{A}, \mathscr{A}'\}$ be a section on l at $\{A, A'\}$ and $\{B, B'\}$. Suppose $\{A, A'\} \neq \{B, B'\}$. First, consider when all of A, A', B, B' are distinct. We shall now show that each of the three possibilities $\mathbf{S}(AB', A'B)$, $\mathbf{S}(AB, A'B')$, and $\mathbf{S}(AA', BB')$ leads to a contradiction.

CASE 1. Let $\mathbf{S}(AB'\ A'B)$. Let $Y \circ l$ such that $\mathbf{S}(AB, B'Y)$. Then $Y \in \mathscr{A}$ or $Y \in \mathscr{A}'$. Consider, first $Y \in \mathscr{A}$. Let $X \in \mathscr{A}'$ such that $X \neq A$, B, A', B' (see Figure 9.5). Then $\mathbf{S}(BB', YX)$ and $\mathbf{S}(AA', YX)$. Now $\mathbf{S}(B'Y, BA)$ and $\mathbf{S}(B'B, YX)$ implies $\mathbf{S}(B'B, AX)$. Further, $\mathbf{S}(B'Y, XA)$ follows. Hence $\mathbf{S}(B'B, AX)$ and $\mathbf{S}(B'A, BA')$ implies $\mathbf{S}(B'A, XA')$. In addition $\mathbf{S}(XY, AA')$ and $\mathbf{S}(XA, YB')$ implies $\mathbf{S}(XA, A'B)$ contrary to

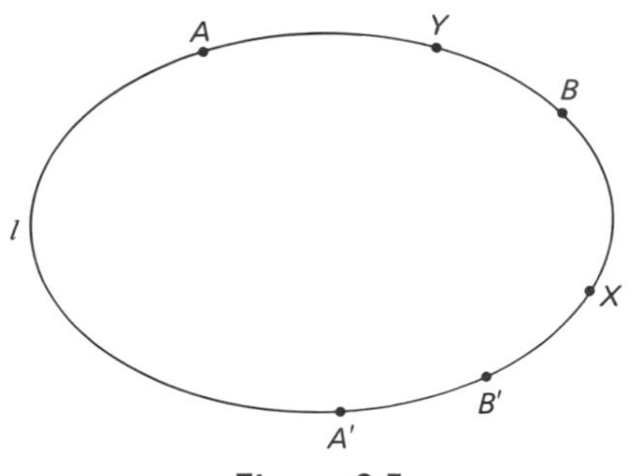

Figure 9.5

$S(B'A, XA')$. Thus $Y \notin \mathscr{A}$. In a similar way, the hypothesis $Y \in \mathscr{A}'$ leads to a contradiction. Therefore $S(AB', A'B)$ cannot hold.

CASE 2. Let $S(AB, A'B')$. This is similar to Case 1.

CASE 3. Let $S(AA', BB')$. Let $Y \circ l$ such that $S(AB', A'Y)$. Then $S(AB', YB)$ and $S(AA', BY)$. Suppose $Y \in \mathscr{A}$. Then $A \in \mathscr{A}$; otherwise, if $A \in \mathscr{A}'$, $S(BB', AY)$ contrary to $S(AB', YB)$. Further, $B \in \mathscr{A}'$ since $S(AA', BY)$ and $Y \in \mathscr{A}$. Hence $B' \in \mathscr{A}$ since $S(AA', BB')$. Now $A' \in \mathscr{A}'$ since $S(BB', AA')$ and $A \in \mathscr{A}$. Let $X \circ l$ such that $S(AB, A'X)$ (see Figure 9.6). Then $S(AA', BB')$ and $S(AB, A'X)$, so that $S(AA', B'X)$. Hence $X \in \mathscr{A}'$. In addition, $S(BB', AA')$ and $S(BA, XA')$, so that $S(BB', XA')$. Hence $X \in \mathscr{A}$. This is a contradiction. Thus $Y \notin \mathscr{A}$. Similarly, $Y \in \mathscr{A}'$ leads to a contradiction. Therefore $S(AA', BB')$ cannot hold. Thus not all of A, A', B, B' can be distinct.

Finally, consider when one equality holds among A, A', B, B', say $A' = B'$. (Recall that $A \neq A'$ and $B \neq B'$.) Then $A \neq B$. That this leads

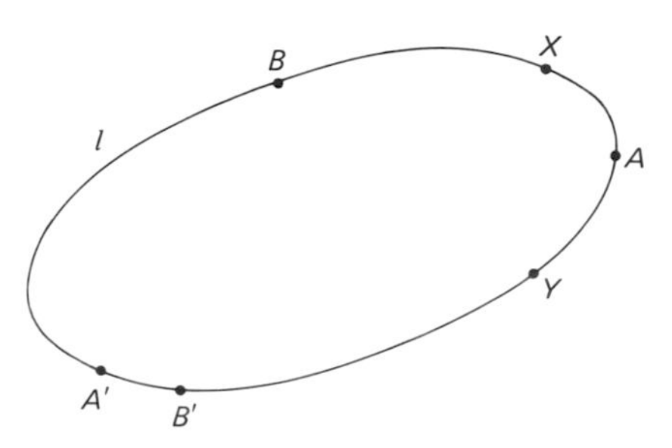

Figure 9.6

to a contradiction is left as an exercise. Thus we must have $\{A, A'\} = \{B, B'\}$. This completes the proof.

We are, now, in a position to state our Axiom of Completeness.

AXIOM C. Every section on a line is at some pair of points on the line.

A separated pappian projective plane satisfying Axiom C will be said to be *complete*. In the remainder of this section, let $(\mathscr{P}, \mathscr{L}, \circ, \mathrm{S})$ be a complete separated pappian projective plane.

Let (Z, U, J, K) be a coordinate system and $(\mathscr{F}, +, \cdot, <)$ its associated ordered field. We shall call $\{\mathscr{H}, \mathscr{H}'\}$ a *cut in* $\mathscr{F}$ if $\mathscr{H}, \mathscr{H}' \subset \mathscr{F}$ such that

(a) $\mathscr{H}, \mathscr{H}' \neq \varnothing$;
(b) $\mathscr{F} = \mathscr{H} \cup \mathscr{H}'$;
(c) if $X \in \mathscr{H}$ and $X' \in \mathscr{H}'$, then $X < X'$.

THEOREM 9.14. Let (Z, U, J, K) be a coordinate system and $(\mathscr{F}, +, \cdot, <)$ its associated ordered field. If $\{\mathscr{H}, \mathscr{H}'\}$ is a cut in $\mathscr{F}$, then there exists $H \in \mathscr{F}$ such that $X \in \mathscr{H}$ and $X' \in \mathscr{H}'$ implies $X \leqq H \leqq X'$.

An ordered field satisfying the property stated in the theorem is said to be *complete*. Thus the ordered field $(\mathscr{F}, +, \cdot, <)$ is complete.

PROOF OF THEOREM 9.14. Let (Z, U, J, K) be a coordinate system and $(\mathscr{F}, +, \cdot, <)$ its associate ordered field with $l = Z \vee U$. Now let $\{\mathscr{H}, \mathscr{H}'\}$ be a cut in $\mathscr{F}$. Define $\mathscr{A} = \mathscr{H}$ and $\mathscr{A}' = \mathscr{H}' \cup \{I\}$. It is left as an exercise to show that there are two points in each of $\mathscr{A}, \mathscr{A}'$ and that $\mathscr{A} \cup \mathscr{A}' = P(l)$. To show that $\{\mathscr{A}, \mathscr{A}'\}$ is a section on l, let $X, Y \in \mathscr{A}$, $X \neq Y$ and $X', Y' \in \mathscr{A}'$, $X' \neq Y'$. Without loss of generality, let $X < Y$.

CASE 1. Let $Y' = I$. Then $X < Y < X'$, so that $Z < Y - X < X' - X$. Hence by Theorem 9.11, $\mathrm{S}(Z(X' - X), (Y - X)I)$. Thus $\mathrm{S}(XX', YI)$, that is, $\mathrm{S}(XX', YY')$.

CASE 2. Let $Y' \neq I$. Without loss of generality, let $X' < Y'$. Then $X < Y < X' < Y'$, so that $Z < Y - X < X' - X < Y' - X$. Hence by Theorem 9.12, $\mathbf{S}(Z(X' - X), (Y - X)(Y' - X))$. Thus $\mathbf{S}(XX', YY')$.

Therefore $\{\mathscr{A}, \mathscr{A}'\}$ is a section on l. Hence by Axiom C and Theorem 9.13, $\{\mathscr{A}, \mathscr{A}'\}$ is at some unique point pair $\{A, A'\}$ on l. Suppose $A, A' \neq I$. Let $A < A'$ without loss of generality. Let $X \in \mathscr{A}$ such that $X \neq A, A'$. Then $A' \not< X$; otherwise, if $A < A' < X$, we have by Theorem 9.12, $\mathbf{S}(AX, A'I)$ contrary to $\mathbf{S}(AA', XI)$. Further, $X \not< A$; otherwise, if $X < A < A'$, $\mathbf{S}(XA', AI)$ contrary to $\mathbf{S}(AA', XI)$. Thus $A < X < A'$. Define $Y = A - (X - A)$. Then $Y < A < X$. Hence $Y \in \mathscr{A}$; otherwise, if $Y \in \mathscr{A}'$, $X < Y$, a contradiction. Now $Y < A < A'$, so that $\mathbf{S}(YA', AI)$ contrary to $\mathbf{S}(AA', YI)$. Thus $A, A' \neq I$ cannot hold. Let $A' = I$ and define $H = A$. Now let $V \in \mathscr{H}$, $V \neq H$ and $V' \in \mathscr{H}'$ $V' \neq H$. Then $V \in \mathscr{A}$ and $V' \in \mathscr{A}'$. Hence $\mathbf{S}(HI, VV')$, that is, $\mathbf{S}(VV', HI)$. Further, $V < V'$. Thus by Theorem 9.12, $V < H < V'$. This completes the proof.

As stated at the beginning of this section, we are assuming that by the *system of real numbers* we mean a complete ordered field. (The use of the definite article "the" in this instance will be discussed in Section 9.5.) Thus if we apply Theorem 6.17, the coordinatization theorem, to our pappian projective plane $(\mathscr{P}, \mathscr{L}, \circ)$, we shall be using the real numbers in the construction of an algebraic incidence basis isomorphic to $(\mathscr{P}, \mathscr{L}, \circ)$. This brings us back to Basis 3.5. We continue the discussion of the relationship between a complete separated pappian projective plane and Basis 3.5 in the next section.

EXERCISES

9.12. For $\{\mathscr{A}, \mathscr{A}'\}$ a section on some line, show that $\mathscr{A} \cap \mathscr{A}' = \varnothing$ and that $\mathscr{A}, \mathscr{A}'$ are not finite.

9.13. Complete the proof of Theorem 9.13 for the case when $A' = B'$.

9.14. In the proof of Theorem 9.14, show that there are two points in each of $\mathscr{A}, \mathscr{A}'$ and that $\mathscr{A} \cup \mathscr{A}' = P(l)$.

9.4. SEPARATION FOR BASIS 3.5

We shall introduce a separation relation S for Basis 3.5 and show that the resultant system is a complete, separated, pappian projective plane. Thus we shall establish the consistency of the notion of a complete, separated, pappian projective plane relative, of course, to the consistency of the real numbers. The relative consistency in the reverse order was established in the last section; namely, an instance of the system of real numbers (a complete ordered field) was constructed in a complete, separated, pappian projective plane.

We look, first, at some additional aspects of Basis 3.5, which will enable us to reduce, significantly, the algebraic computations involved in the introduction of our separation relation. In the remainder of this section, let $(\mathscr{P}, \mathscr{L}, \circ)$ be Basis 3.5, which is to be the object of study, and the remaining statements are to be considered in this context.

LEMMA 9.15. Let (x, y, z) and (x', y', z') be distinct points and $[a, b, c]$ a line. Then $(x, y, z), (x', y', z') \circ [a, b, c]$ if and only if

$$[a, b, c] \equiv \left[\begin{vmatrix} y & z \\ y' & z' \end{vmatrix}, -\begin{vmatrix} x & z \\ x' & z' \end{vmatrix}, \begin{vmatrix} x & y \\ x' & y' \end{vmatrix}\right].$$

PROOF. For the "only if" part, let the points be on the line, that is,

$$\begin{aligned} ax + by + cz &= 0, \\ ax' + by' + cz' &= 0. \end{aligned} \tag{1}$$

Define $x'' = yz' - y'z$, $y'' = xz' - x'z$ and $z'' = xy' - x'y$. Without loss of generality there are three essential cases to be considered.

CASE 1. Let $b = c = 0$. Then $a \neq 0$ and $x = x' = 0$. Suppose $x'' = 0$. Then $yz' = y'z$. Not both $y = 0$ and $z = 0$. Let $z \neq 0$. Then $y' = y(z'/z)$. Further, $z' \neq 0$; otherwise, if $z' = 0$, we have $y' = 0$ contrary to $(x', y', z') \neq (0, 0, 0)$. Now $x' = xk, y' = yk, z' = zk$ and $k \neq 0$ for $k = z'/z$ contrary to our points being distinct. The hypothesis $y \neq 0$ leads, similarly, to a contradiction. Thus $x'' \neq 0$. Now for $j = x''/a$, we have $x'' = ja$, $-y'' = jb$, $z'' = jc$, and $j \neq 0$.

CASE 2. Let $a, b \neq 0$ and $c = 0$. Then (1) becomes

$$ax + by = 0,$$
$$ax' + by' = 0.$$

Combining these statements to eliminate the terms in a, we obtain $b(xy' - x'y) = 0$. Hence $xy' - x'y = 0$, and therefore $z'' = 0$. Suppose $x'' = 0$ and $y'' = 0$. Then $yz' = y'z$ and $xz' = x'z$. None of x, y, x', y' can be 0; otherwise, if one is 0, all are 0 contrary to (x, y, z) and (x', y', z') being distinct. Now if $z = 0$, then $z' = 0$, and $x' = xk, y' = yk$, $z' = zk$ and $k \neq 0$ for $k = y'/y$, a contradiction. Further, if $z \neq 0$, $k = z'/z$ leads to the same contradiction. Hence $x'' \neq 0$ or $y'' \neq 0$. Without loss of generality, let $x'' \neq 0$. Now for $j = x''/a$, $x'' = ja$, $-y'' = x'z - xz' = (-(b/a)y')z - (-(b/a)y')z' = ((yz' - y'z)/a)b = jb$, $z'' = jc$, and $j \neq 0$.

CASE 3. Let $a, b, c \neq 0$. Without loss of generality, let $x \neq 0$. Combine the statements in (1) to eliminate the terms in a and obtain $-x'(by + cz) + x(by' + cz') = 0$, that is, $bz'' = c(-y'')$. If we similarly eliminate the terms in b, we have $az'' = cx''$. Suppose, $z'' = 0$. Then $y'' = 0$, so that $xy' = x'y$ and $xz' = x'z$. In addition, $x' \neq 0$; otherwise, if $x' = 0$, we have $y' = z' = 0$, a contradiction. Now for $k = x'/x$, $x' = xk$, $y' = yk$, $z' = zk$, and $k \neq 0$, a contradiction. Hence $z'' \neq 0$. Let $j = z''/c$. Then $x'' = ja$, $-y'' = jb$, $z'' = jc$, and $j \neq 0$.

The converse is left as an exercise. This completes the proof.

We have the following as an immediate consequence of Lemma 5.4 in Chapter 5: If $P \equiv (x, y, z)$, $P' = (x', y', z')$, $P'' = (x'', y'', z'')$ are collinear on line l and $P' \not\equiv P''$, then there exist j and k, not both 0, such that $x = jx' + kx''$, $y = jy' + ky''$, $z = jz' + kz''$. Moreover, the pair (j, k) is unique up to a nonzero multiplier. Thus we have, in effect, a "parametric representation" of the points on l modulo the "base" points P' and P'' with "parameters" j and k. We shall employ this representation in what follows. As a notational convenience we shall write (x_i) for (x_1, x_2, x_3) and $[m_i]$ for $[m_1, m_2, m_3]$.

LEMMA 9.16. Let $P \not o\, l$ and let $(x_i), (y_i) \circ l$ such that $(x_i) \not\equiv (y_i)$ and $[m_i], [n_i] \circ P$ such that $[m_i] \not\equiv [n_i]$. Then there exist real numbers a, b, c, d such that

(a) $ad - bc \neq 0$;

(b) for $Q = (jx_i + ky_i)$ with $j^2 + k^2 \neq 0$ and $q = [j'm_i + k'n_i]$ with $j'^2 + k'^2 \neq 0$ such that $Q \circ q$,

$$ej' = aj + bk,$$

$$ek' = cj + dk$$

for some nonzero real number e.

PROOF. Let Q and q be as stated in the hypothesis. Then $Q \circ q$ means $\sum(j'm_i + k'n_i)(jx_i + ky_i) = 0$. (All sums in the $\sum$-notation will be for the index values $i = 1, 2, 3$ and the limits of summation will be omitted from the notation.) Hence

$$j'[(\sum m_i x_i)j + (\sum m_i y_i)k] + k'[(\sum n_i x_i)j + (\sum n_i y_i)k] = 0.$$

Thus $j'(cj + dk) - k'(aj + bk) = 0$ if we define $c = \sum m_i x_i$, $d = \sum m_i y_i$, $a = -\sum n_i x_i$ and $b = -\sum n_i y_i$. It is left as an exercise to show the existence of the desired real number e. To show that $ad - bc \neq 0$, suppose the contrary. Then

$$(-\sum n_i x_i)(\sum m_i y_i) - (-\sum n_i y_i)(\sum m_i x_i) = 0.$$

This reduces to

$$\begin{vmatrix} x_2 & x_3 \\ y_2 & y_3 \end{vmatrix} \cdot \begin{vmatrix} m_2 & m_3 \\ n_2 & n_3 \end{vmatrix} + \begin{vmatrix} x_1 & x_3 \\ y_1 & y_3 \end{vmatrix} \cdot \begin{vmatrix} m_1 & m_3 \\ n_1 & n_3 \end{vmatrix} + \begin{vmatrix} x_1 & x_2 \\ y_1 & y_2 \end{vmatrix} \cdot \begin{vmatrix} m_1 & m_2 \\ n_1 & n_2 \end{vmatrix} = 0.$$

Hence by Lemma 9.15,

$$R = \left(\begin{vmatrix} m_2 & m_3 \\ n_2 & n_3 \end{vmatrix}, -\begin{vmatrix} m_1 & m_3 \\ n_1 & n_3 \end{vmatrix}, \begin{vmatrix} m_1 & m_2 \\ n_1 & n_2 \end{vmatrix}\right)$$

is collinear with (x_i) and (y_i). By the dual of Lemma 9.15, R is on $[m_i]$ and $[n_i]$. Therefore $P \circ l$ contrary to hypothesis. Thus $ad - bc \neq 0$. This completes the proof.

THEOREM 9.17. Let π be a projectivity between lines l and l' and let $(x_i), (y_i) \circ l$ such that $(x_i) \not\equiv (y_i)$ and $(x_i'), (y_i') \circ l'$ such that $(x_i') \not\equiv (y_i')$. Then there exist real numbers a, b, c, d such that

(a) $ad - bc \neq 0$;

(b) for $P = (jx_i + ky_i)$ with $j^2 + k^2 \neq 0$ and $\pi P = (j'x_i' + k'y_i')$,

$$ej' = aj + bk,$$
$$ek' = cj + dk$$

for some nonzero real number e.

The proof is left as an exercise. (Note that a perspectivity between two lines can be described algebraically by combining the equations from Lemma 9.16 and the equations from the dual of the lemma.)

COROLLARY 9.18. In Theorem 9.17, if $(x_i') = \pi(x_i)$ and $(y_i') = \pi(y_i)$, then $b = c = 0$.

The proof is left as an exercise.

Let (x_i), $(y_i) \circ l$ such that $(x_i) \not\equiv (y_i)$. Now let A, B, C, D be distinct points on l such that $A \equiv (a_1 x_i + a_2 y_i)$, $B \equiv (b_1 x_i + b_2 y_i)$, $C \equiv (c_1 x_i + c_2 y_i)$, and $D \equiv (d_1 x_i + d_2 y_i)$. Define the *cross-ratio* of (A, B, C, D), written $R(A, B, C, D)$, by

$$R(A, B, C, D) = \frac{\begin{vmatrix} a_1 & a_2 \\ c_1 & c_2 \end{vmatrix}}{\begin{vmatrix} b_1 & b_2 \\ c_1 & c_2 \end{vmatrix}} \div \frac{\begin{vmatrix} a_1 & a_2 \\ d_1 & d_2 \end{vmatrix}}{\begin{vmatrix} b_1 & b_2 \\ d_1 & d_2 \end{vmatrix}}$$

The ensuing discussion deals with the sense in which R is well defined. First, none of the determinant values in the defining expression is 0 since A, B, C, D are distinct (whence do not have pairs of "parameter" values which are related by a nonzero multiplier). Hence we also have $R(A, B, C, D) \neq 0$. Secondly, the value of the defining expression is independent of the particular choice of "parameter" values used in the representation of A, B, C, and D. To see this, let j, k, m, $n \neq 0$. Then $A \equiv ((ja_1)x_i + (ja_2)y_i)$, $B \equiv ((kb_1)x_i + (kb_2)y_i)$, $C \equiv ((mc_1)x_i + (mc_2)y_i)$, and $D \equiv ((nd_1)x_i + (nd_2)y_i)$, and

$$\frac{\begin{vmatrix} ja_1 & ja_2 \\ mc_1 & mc_2 \end{vmatrix}}{\begin{vmatrix} kb_1 & kb_2 \\ mc_1 & mc_2 \end{vmatrix}} \div \frac{\begin{vmatrix} ja_1 & ja_2 \\ nd_1 & nd_2 \end{vmatrix}}{\begin{vmatrix} kb_1 & kb_2 \\ nd_1 & nd_2 \end{vmatrix}} = \frac{jm\begin{vmatrix} a_1 & a_2 \\ c_1 & c_2 \end{vmatrix}}{km\begin{vmatrix} b_1 & b_2 \\ c_1 & c_2 \end{vmatrix}} \div \frac{jn\begin{vmatrix} a_1 & a_2 \\ d_1 & d_2 \end{vmatrix}}{kn\begin{vmatrix} b_1 & b_2 \\ d_1 & d_2 \end{vmatrix}} = \frac{\begin{vmatrix} a_1 & a_2 \\ c_1 & c_2 \end{vmatrix}}{\begin{vmatrix} b_1 & b_2 \\ c_1 & c_2 \end{vmatrix}} \div \frac{\begin{vmatrix} a_1 & a_2 \\ d_1 & d_2 \end{vmatrix}}{\begin{vmatrix} b_1 & b_2 \\ d_1 & d_2 \end{vmatrix}}.$$

Finally, the value of the defining expression is independent of the "base" points (x_i) and (y_i) used in the "parametric representation." To realize this, let $(x_i'), (y_i') \circ l$ such that $(x_i') \not\equiv (y_i')$ and let $A \equiv (a_1' x_i' + a_2' y_i')$, $B \equiv (b_1' x_i' + b_2' y_i')$, $C \equiv (c_1' x_i' + c_2' y_i')$, and $D \equiv (d_1' x_i' + d_2' y_i')$. Now let π be a self projectivity of l such that $(x_i') = \pi(x_i)$ and $(y_i') = \pi(y_i)$. By Corollary 9.18, there exist nonzero real numbers m, n, e_a, e_b, e_c, e_d such that

$$\begin{array}{llll} e_a a_1' = m a_1, & e_b b_1' = m b_1, & e_c c_1' = m c_1, & e_d d_1' = m d_1, \\ e_a a_2' = n a_2, & e_b b_2' = n b_2, & e_b c_2' = n c_2, & e_d d_2' = n d_2. \end{array}$$

Now

$$\begin{vmatrix} a_1' & a_2' \\ c_1' & c_2' \end{vmatrix} = \begin{vmatrix} \frac{m}{e_a} a_1 & \frac{n}{e_a} a_2 \\ \frac{m}{e_c} c_1 & \frac{n}{e_c} c_2 \end{vmatrix} = \frac{mn}{e_a e_c} \cdot \begin{vmatrix} a_1 & a_2 \\ c_1 & c_2 \end{vmatrix}.$$

We have a similar property for each of the determinants. Hence

$$\frac{\begin{vmatrix} a_1' & a_2' \\ c_1' & c_2' \end{vmatrix}}{\begin{vmatrix} b_1' & b_2' \\ c_1' & c_2' \end{vmatrix}} \div \frac{\begin{vmatrix} a_1' & a_2' \\ d_1' & d_2' \end{vmatrix}}{\begin{vmatrix} b_1' & b_2' \\ d_1' & d_2' \end{vmatrix}} = \frac{\frac{mn}{e_a e_c}\begin{vmatrix} a_1 & a_2 \\ c_1 & c_2 \end{vmatrix}}{\frac{mn}{e_b e_c}\begin{vmatrix} b_1 & b_2 \\ c_1 & c_2 \end{vmatrix}} \div \frac{\frac{mn}{e_a e_d}\begin{vmatrix} a_1 & a_2 \\ d_1 & d_2 \end{vmatrix}}{\frac{mn}{e_b e_d}\begin{vmatrix} b_1 & b_2 \\ d_1 & d_2 \end{vmatrix}} = \frac{\begin{vmatrix} a_1 & a_2 \\ c_1 & c_2 \end{vmatrix}}{\begin{vmatrix} b_1 & b_2 \\ c_1 & c_2 \end{vmatrix}} \div \frac{\begin{vmatrix} a_1 & a_2 \\ d_1 & d_2 \end{vmatrix}}{\begin{vmatrix} b_1 & b_2 \\ d_1 & d_2 \end{vmatrix}}.$$

Thus R is well defined.

LEMMA 9.19. Let A, B, C, D, E be distinct collinear points. Then

(a) $R(A, B, C, D) = [R(A, B, D, C)]^{-1}$;
(b) $R(A, B, C, D) + R(A, C, B, D) = 1$;
(c) $R(A, B, C, D) \neq 1$;
(d) $R(A, C, B, E) \cdot R(A, C, E, D) = R(A, C, B, D)$;
(e) for π a projectivity between $A \vee B$ and some line l

$$R(\pi A, \pi B, \pi C, \pi D) = R(A, B, C, D).$$

PROOF. The proof of parts (a)–(d) is left as an exercise. For part (e), let π be a projectivity. Then let $A = (x_i)$, $B = (y_i)$ be "base" points on

$A \vee B$ and $\pi A = (x_i')$, $\pi B = (y_i')$ be "base" points on $(\pi A) \vee (\pi B)$. Finally, let $C = (c_1 x_i + c_2 y_i)$, $D = (d_1 x_i + d_2 y_i)$, $\pi C = (c_1' x_i' + c_2' y_i')$, and $\pi D = (d_1' x_i' + d_2' y_i')$. By Corollary 9.18, there exist m, n, e_a $e_b, e_c, e_d \neq 0$ such that

$$\begin{array}{llll} e_a \cdot 1 = m \cdot 1, & e_b \cdot 0 = m \cdot 0, & e_c \cdot c_1' = m \cdot c_1, & e_d \cdot d_1' = m \cdot d_1, \\ e_a \cdot 0 = n \cdot 0, & e_b \cdot 1 = n \cdot 1, & e_c \cdot c_2' = n \cdot c_2, & e_d \cdot d_2' = n \cdot d_2. \end{array}$$

Now

$$\begin{aligned} R(\pi A, \pi B, \pi C, \pi D) &= \frac{\begin{vmatrix} 1 & 0 \\ c_1' & c_2' \end{vmatrix}}{\begin{vmatrix} 0 & 1 \\ c_1' & c_2' \end{vmatrix}} \div \frac{\begin{vmatrix} 1 & 0 \\ d_1' & d_2' \end{vmatrix}}{\begin{vmatrix} 0 & 1 \\ d_1' & d_2' \end{vmatrix}} \\ &= \frac{\dfrac{mn}{e_a e_c} \cdot \begin{vmatrix} 1 & 0 \\ c_1 & c_2 \end{vmatrix}}{\dfrac{mn}{e_b e_c} \cdot \begin{vmatrix} 0 & 1 \\ c_1 & c_2 \end{vmatrix}} \div \frac{\dfrac{mn}{e_a e_d} \cdot \begin{vmatrix} 1 & 0 \\ d_1 & d_2 \end{vmatrix}}{\dfrac{mn}{e_b e_d} \cdot \begin{vmatrix} 0 & 1 \\ d_1 & d_2 \end{vmatrix}} \\ &= R(A, B, C, D). \end{aligned}$$

Now we are in a position to define our separation relation $\mathbf{S}$ as

$\mathbf{S} = \{(A, B, C, D)$: A, B, C, D are distinct collinear points such that $R(A, B, C, D) < 0\}$.

THEOREM 9.20. The system $(\mathscr{P}, \mathscr{L}, \circ, \mathbf{S})$ is a complete, separated pappian projective plane.

PROOF. We already know that $(\mathscr{P}, \mathscr{L}, \circ)$ is a pappian projective plane. It is immediate that $\mathbf{S} \neq \varnothing$. The proof of Axioms S_1–S_3 and S_5 is left as an exercise. (Note that for Axiom S_3, if A, B, C, D are distinct collinear points, then $R(A, B, C, D) \cdot R(A, C, D, B) \cdot R(A, D, B, C) = -1$.) For Axiom S_4, let A, B, C, D, E be distinct collinear points such that $\mathbf{S}(AB, CD)$ and $\mathbf{S}(AC, BE)$. Then $R(A, B, C, D), R(A, C, B, E) < 0$. By part (b) of Lemma 9.19, $R(A, C, B, D) = 1 - R(A, B, C, D) > 0$. Hence by part (d), $R(A, C, B, E) \cdot R(A, C, E, D) = R(A, C, B, D) > 0$, so that $R(A, C, E, D) < 0$. Now by part (a),

$$R(A, C, D, E) = [R(A, C, E, D)]^{-1} < 0.$$

Hence $S(AC, DE)$, that is, Axiom S_4 is satisfied. Thus $(\mathscr{P}, \mathscr{L}, \circ, S)$ is a separated projective plane.

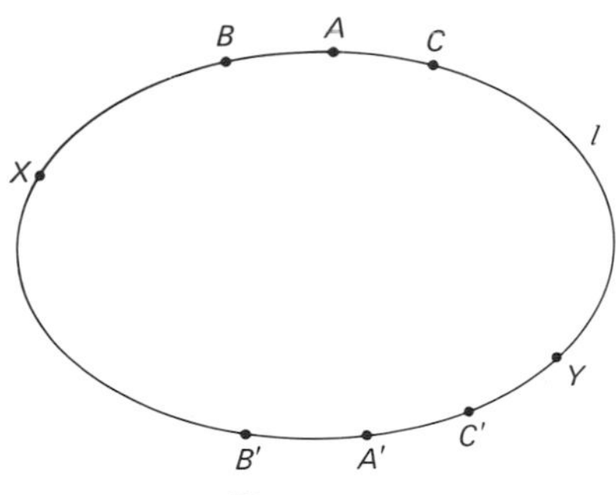

Figure 9.7

Finally, for Axiom C, let $\{\mathscr{A}, \mathscr{A}'\}$ be a section on some line l. Let $X, B, B' \in \mathscr{A}$, $Y, C, C' \in \mathscr{A}'$ such that $S(XY, BB')$, $S(XY, CC')$ (see Figure 9.7). Let $X = (x_i)$ and $Y = (y_i)$. Then there exist b, c, b' $c' \neq 0$ such that $B \equiv (x_i + by_i)$, $C \equiv (x_i + cy_i)$, $B' \equiv (x_i + b'y_i)$, and $C' \equiv (x_i + c'y_i)$. In addition,

$$R(X, C, B, Y) = \frac{\begin{vmatrix} 1 & 0 \\ 1 & b \end{vmatrix}}{\begin{vmatrix} 1 & c \\ 1 & b \end{vmatrix}} \div \frac{\begin{vmatrix} 1 & 0 \\ 0 & 1 \end{vmatrix}}{\begin{vmatrix} 1 & c \\ 0 & 1 \end{vmatrix}} = \frac{b}{b - c},$$

$$R(X, Y, B, B') = \frac{\begin{vmatrix} 1 & 0 \\ 1 & b \end{vmatrix}}{\begin{vmatrix} 0 & 1 \\ 1 & b \end{vmatrix}} \div \frac{\begin{vmatrix} 1 & 0 \\ 1 & b' \end{vmatrix}}{\begin{vmatrix} 0 & 1 \\ 1 & b' \end{vmatrix}} = \frac{b}{b'},$$

and

$$R(X, C', B', Y) = \frac{\begin{vmatrix} 1 & 0 \\ 1 & b' \end{vmatrix}}{\begin{vmatrix} 1 & c' \\ 1 & b' \end{vmatrix}} \div \frac{\begin{vmatrix} 1 & 0 \\ 0 & 1 \end{vmatrix}}{\begin{vmatrix} 1 & c' \\ 0 & 1 \end{vmatrix}} = \frac{b'}{b' - c'}.$$

Without loss of generality, let $b > 0$. Then $c' < b' < 0 < b < c$. Define $\mathscr{I} = \{t : 0 < t < c$ and for $T = (x_i + ty_i)$, $T \in \mathscr{A}$ and $S(XC, TY)\}$,

$\mathscr{H} = \{r : r \leqq t \text{ for some } t \in \mathscr{S}\}$ and $\mathscr{H}' = \{s : s > t \text{ for every } t \in \mathscr{S}\}$. Then (a) $b \in \mathscr{H}$ and $c \in \mathscr{H}'$: (b) $\mathscr{H} \cup \mathscr{H}'$ is the set of real numbers; (c) if $r \in \mathscr{H}$ and $s \in \mathscr{H}'$, $r < s$. Hence $\{\mathscr{H}, \mathscr{H}'\}$ is a cut, so that there exists h such that $r \leqq h \leqq s$ for $r \in \mathscr{H}$, $s \in \mathscr{H}'$. For $0 < r < h < s$, $R = (x_i + ry_i)$ and $S = (x_i + sy_i)$, we have S(XS, RY). Similarly, we can find h' such that $c' \leqq h' \leqq b'$, and for $s < h' < r < 0$, $R = (x_i + ry_i)$ and $S = (x_i + sy_i)$, we have S(XS, RY). Define $A = (x_i + hy_i)$ and $A' = (x_i + h'y_i)$. It is left as an exercise to show that $\{\mathscr{A}, \mathscr{A}'\}$ is at $\{A, A'\}$. Thus $(\mathscr{P}, \mathscr{L}, \circ, \text{S})$ is a complete, separated, pappian projective plane.

EXERCISES

9.15. Prove the converse part of Theorem 9.15.
9.16. Verify the statements made in the paragraph preceding Lemma 9.16 concerning the "parametric representation."
9.17. In the proof of Lemma 9.16, show the existence of the real number e with the desired properties.
9.18. Prove Theorem 9.17.
9.19. For the converse of Theorem 9.17, let the transformation π between l and l' be given algebraically by the equations and the "parametric" representation of the points on l and l' (select $e = 1$). Show that π is a projectivity.
9.20. Prove Corollary 9.18.
9.21. Prove parts (a)–(d) of Lemma 9.19.
9.22. Prove Axioms S_1–S_3 and S_5 in the proof of Theorem 9.20.
9.23. For Axiom C, in the proof of Theorem 9.20, show that X, Y, B, B', C, C' exist with the stated properties and that $\{\mathscr{A}, \mathscr{A}'\}$ is at $\{A, A'\}$.

9.5. THE REAL PROJECTIVE PLANE

A complete, separated, pappian projective plane will be called a *real projective plane*. We study, now, an isomorphism property of real projective planes, which suggests speaking of *the* real projective plane.

Let $(\mathscr{P}_1, \mathscr{L}_1, \circ_1, \text{S}_1)$ and $(\mathscr{P}_2, \mathscr{L}_2, \circ_2, \text{S}_2)$ be incidence bases with separation relations. We say $(\mathscr{P}_1, \mathscr{L}_1, \circ_1, \text{S}_1)$ and $(\mathscr{P}_2, \mathscr{L}_2, \circ_2, \text{S}_2)$

are *isomorphic* if there exists an isomorphism (π, λ) between $(\mathscr{P}_1, \mathscr{L}_1, \circ_1)$ and $(\mathscr{P}_2, \mathscr{L}_2, \circ_2)$ such that for $A, B, C, D \in \mathscr{P}_1$, $\mathbf{S}_1(AB, CD)$ if and only if $\mathbf{S}_2((\pi A)(\pi B), (\pi C)(\pi D))$.

THEOREM 9.21. A real projective plane is isomorphic to an algebraic incidence basis with a separation relation, the incidence basis constructed from some complete ordered field.

PROOF. Let $(\mathscr{P}, \mathscr{L}, \circ, \mathbf{S})$ be a real projective plane. Let (Z, U, J, K) be a coordinate system with auxiliary elements l, j, k, m, and I in $(\mathscr{P}, \mathscr{L}, \circ)$, and $(\mathscr{F}, +, \cdot, <)$ the complete ordered field associated with (Z, U, J, K). Finally, let $(\mathscr{P}_1, \mathscr{L}_1, \circ_1, \mathbf{S}_1)$ be the real projective plane, where $(\mathscr{P}_1, \mathscr{L}_1, \circ_1)$ is the algebraic incidence basis with elements from $(\mathscr{F}, +, \cdot)$ and $\mathbf{S}_1$ is defined as in the last section. Consider the isomorphism (π, λ) between $(\mathscr{P}, \mathscr{L}, \circ)$ and $(\mathscr{P}_1, \mathscr{L}_1, \circ_1)$ introduced in the proof of Theorem 6.17 and restated here.

For $P \in \mathscr{P}$,

$$\pi P = \begin{cases} (S, T, U) & \text{if } P \not\circ m \text{ and } P : (S, T), \\ (U, R, Z) & \text{if } P \circ m, P \neq K \text{ and } P \text{ is on the line with equation } Y = R \cdot X, \\ (Z, U, Z) & \text{if } P = K. \end{cases}$$

For $n \in \mathscr{L}$,

$$\lambda n = \begin{cases} [-A, U, -B] & \text{if } n \not\circ K \text{ and } n \text{ has equation } Y = A \cdot X + B, \\ [U, Z, -C] & \text{if } n \circ K, n \neq m, \text{ and } n \text{ has equation } X = C, \\ [Z, Z, U] & \text{if } n = m. \end{cases}$$

For the condition of the isomorphism relating $\mathbf{S}$ and $\mathbf{S}_1$, it will be sufficient to consider distinct and collinear $A, B, C, D \in \mathscr{P}$ such that $A = Z$, $C = U$, $D = I$, and $B \circ l$. (Why?) Let $B \circ l$. Then $\pi Z = (Z, Z, U)$, $\pi B = (B, B, U)$, $\pi U = (U, U, U)$, and $\pi I = (U, U, Z)$. Using πZ and πI as "base" points we have

$$R(\pi Z, \pi B, \pi U, \pi I) = \frac{\begin{vmatrix} U & Z \\ U & U \end{vmatrix}}{\begin{vmatrix} U & B \\ U & U \end{vmatrix}} \div \frac{\begin{vmatrix} U & Z \\ Z & U \end{vmatrix}}{\begin{vmatrix} U & B \\ Z & U \end{vmatrix}} = \frac{U}{U - B}.$$

Now the following statements are equivalent:

(a) $\mathbf{S}(ZB, UI)$;

(b) $U < B$;

(c) $\dfrac{U}{U - B} < Z$;

(d) $\mathbf{S}_1((\pi Z)(\pi B), (\pi U)(\pi I))$.

This completes the proof.

We state an additional property of complete ordered fields and, with the last theorem, apply this property to real projective planes. We shall say that ordered fields $(\mathscr{F}_1, +_1, \cdot_1, <_1)$ and $(\mathscr{F}_2, +_2, \cdot_2, <_2)$ are *isomorphic* if there exists a one-to-one correspondence σ between $\mathscr{F}_1$ and $\mathscr{F}_2$ such that for $x, y \in \mathscr{F}_1$,

(a) $\sigma(x +_1 y) = \sigma(x) +_2 \sigma(y)$;

(b) $\sigma(x \cdot_1 y) = \sigma(x) \cdot_2 \sigma(y)$;

(c) $x <_1 y$ if and only if $\sigma(x) <_2 \sigma(y)$.

The following is the property of interest to us, which is a result from the theory of real numbers and which we shall accept as being given (see Eves and Newsom [9]).

Any two complete ordered fields are isomorphic. This is the justification for speaking of *the* system of real numbers. Moreover, we say that the set of axioms for a complete ordered field is *categorical.* In general, we define the notion of *categoricity* for any set of axioms (stated for a basis for which we have defined isomorphism) when such a condition of isomorphism is met.

THEOREM 9.22. The set of axioms for a real projective plane is categorical.

This follows from our last theorem and the assumed result from the theory of real numbers. We are, finally, in a position to speak of *the real projective plane.*

EXERCISES

9.24. In the proof of Theorem 9.21, when proving the isomorphism condition relating **S** and $\mathbf{S}_1$, points A, B, C, D were chosen so that $A = Z$, $C = U$, $D = I$, and $B \circ l$. Show that this is sufficient.
9.25. Have we studied other instances of categorical sets of axioms?

9.6. EUCLIDEAN PLANES

We shall conclude this chapter with an outline of the relationship between the real projective plane and euclidean planes. No attempt will be made to supply all of the details.

Let $(\mathscr{P}_1, \mathscr{L}_1, \circ_1)$ be a euclidean plane imbedded, as an affine plane, into the projective plane $(\mathbf{P}, \mathbf{L}, \circ)$. The latter will be pappian for, indeed, the Theorem of Pappus is a theorem of euclidean geometry [even when extended to include the affine interpretation of the ideal points and line of $(\mathbf{P}, \mathbf{L}, \circ)$]. The euclidean betweenness relation for points can be used to define a separation relation **S** for the projective plane as follows: Let A, B, C, D be distinct euclidean points on a line l such that B is between A and C, and C is between B and D. In addition, let a, b, c, d be lines concurrent on a point O (not on l) and incident with A, B, C, D, respectively (see Figure 9.8). Then we define

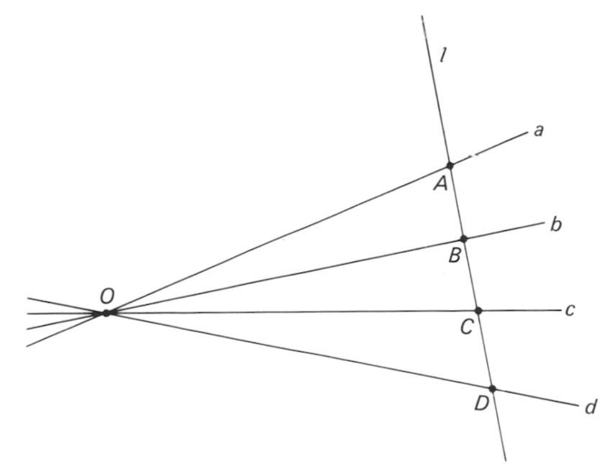

Figure 9.8

(a) $\mathbf{S}(L(A)L(C), L(B)L(D))$;
(b) $\mathbf{S}(L(A)L(C), L(B)L(l))$;
(c) $\mathbf{S}(L(a)L(c), L(b)L(d))$;
(d) $\mathbf{S}(L(a)L(c), L(b)L(l))$.

These four cases represent all of the possible ways we might have four distinct collinear points in the projective plane. Now $(\mathbf{P}, \mathbf{L}, \circ, \mathbf{S})$ can be shown to be the real projective plane. Thus when imbedding a euclidean plane, the resultant projective plane is necessarily the real projective plane if separation is defined as above (see Tuller [17]).

For the sufficiency of the above condition, we must show that any deletion subgeometry of the real projective plane is a euclidean plane where, of course, additional concepts will have been suitably defined. Let $(\mathscr{P}, \mathscr{L}, \circ, \mathbf{S})$ be the real projective plane, $l^* \in \mathscr{L}$ and $(\mathscr{P}', \mathscr{L}', \odot)$ be the deletion subgeometry associated with l^*. We can define a betweenness relation in the affine plane $(\mathscr{P}', \mathscr{L}', \odot)$ as follows: For a triple of distinct points $A, B, C \in \mathscr{P}'$ collinear on a line $l \in \mathscr{L}'$, B is defined to be between A and C if $\mathbf{S}(AC, B(l \wedge l^*))$. Now Hilbert's Axioms of Connection, Parallels, and Order can be shown to hold. This leaves the Axioms of Congruence and Continuity.

To complete the sufficiency discussion, we shall appeal to analytic geometry, both from our projective theory and euclidean geometry. By Theorem 9.21, $(\mathscr{P}, \mathscr{L}, \circ, \mathbf{S})$ is isomorphic to an algebraic incidence basis $(\mathscr{P}_1, \mathscr{L}_1, \circ_1, \mathbf{S}_1)$ with a separation relation, where $(\mathscr{P}_1, \mathscr{L}_1, \circ_1)$ is constructed from the real numbers. In particular, in this coordinatization process, let us choose (Z, U, J, K) so that $l^* = J \vee K$. Then, under the isomorphism, l^* corresponds to $[0, 0, 1] \in \mathscr{L}_1$. (We shall write 0 and 1, respectively, for Z and U, the additive and multiplicative identities.) Recall that $[0, 0, 1]$ was the line deleted from Basis 3.5 to obtain a subgeometry isomorphic to Basis 2.7 (see the discussion following the proof of Theorem 4.4). Designate the deletion subgeometry of $(\mathscr{P}_1, \mathscr{L}_1, \circ_1)$ associated with $[0, 0, 1]$ by $(\mathscr{P}_1', \mathscr{L}_1', \odot_1)$ and Basis 2.7 by $(\mathscr{P}_2, \mathscr{L}_2, \circ_2)$. Now the above isomorphisms induce an isomorphism between $(\mathscr{P}', \mathscr{L}', \odot)$ and $(\mathscr{P}_2, \mathscr{L}_2, \circ_2)$. Moreover, the betweenness relation for the points of $(\mathscr{P}', \mathscr{L}', \odot)$ can be described analytically in terms of this induced isomorphism.

The isomorphism that can be defined between a euclidean plane and Basis 2.7 is based on the choice of a rectangular cartesian coordinate system with equal unit length on the two axes. (We are assuming this to be known.) The measure of the distance between two points and the measure of angles can be described analytically by formulas in terms of this isomorphism. (The measure of angles involves the inverse cosine function.)

Now we use these formulas to define a measure of distance between points and a measure of angles in $(\mathscr{P}', \mathscr{L}', \odot)$ in terms of the isomorphism between $(\mathscr{P}', \mathscr{L}', \odot)$ and $(\mathscr{P}_2, \mathscr{L}_2, \circ_2)$. (See Appendix A for definitions of *angle* and *line segment*.) Congruence of line segments and angles can, then, be defined in terms of these measures so that the remaining axioms of Hilbert can be proved. Thus it is sufficient that $(\mathscr{P}, \mathscr{L}, \circ, \mathrm{S})$ be the real projective plane in order that any deletion subgeometry, with betweenness and congruence defined as in the above, be a euclidean plane. It is of interest to note that the nondeleted elements of the coordinate system (Z, U, J, K) will form a rectangular cartesian coordinate system with equal unit length on the two axes in the euclidean plane.

PROJECTIVE SPACES—PART 1 CHAPTER 10

Let us return to our first set of axioms for a projective plane (restated here for reference) and study an aspect not previously emphasized. Let $(\mathscr{P}, \mathscr{L}, \circ)$ be an incidence basis.

AXIOM P_1. There exist a point P and a line l such that $P \not\circ l$.

AXIOM P_2. Each line has at least three points on it.

AXIOM P_3. If P and Q are distinct points, there exists one and only one line l such that $P \circ l$ and $Q \circ l$.

AXIOM P_4. If l and m are distinct lines, there exists a point P such that $P \circ l$ and $P \circ m$.

Let $P \in \mathscr{P}$ and $l \in \mathscr{L}$ such that $P \not\circ l$. Then define

$$S = \{R \in \mathscr{P} : R \circ P \vee Q \quad \text{for some} \quad Q \circ l\}.$$

Certainly, $S \subset \mathscr{P}$. For the reverse inclusion, let $R \in \mathscr{P}$. If $R = P$, then $R \circ P \vee Q$ where Q is any point on l. If $R \circ l$, then $R \circ P \vee R$. Finally for $R \neq P$ and $R \not\circ l$, define $Q = (P \vee R) \wedge l$, Then $R \circ P \vee Q$ and $Q \circ l$. In all cases, $R \in S$. Hence $\mathscr{P} \subset S$. Thus $S = \mathscr{P}$. (This, incidentally, motivates the term "plane," that is, an object determined by a point and a line, the point not incident with the line.)

In the above argument that $\mathscr{P} \subset S$, a critical use of Axiom P_4 was made in the case when $R \neq P$ and $R \not\circ l$. In fact, it is essentially this axiom that confines all the points of $\mathscr{P}$ to a "plane." It is apparent, then, that in order to have points not in S (that is, not necessarily confined to a single "plane") Axiom P_4 will have to be modified in some manner. To dispense with it entirely would not be satisfactory since we shall want enough structure to be able to speak of "planes" within whatever "space" concept we describe. A workable compromise is the following.

AXIOM. If P, Q, R, T are distinct points such that $P \vee Q$ and $R \vee T$ are on a common point, then $P \vee R$ and $Q \vee T$ are on a common point.

This is generally called "Pasch's Axiom" (after Moritz Pasch, 1843–1930) (see Figure 10.1). It guarantees the intersection of lines only if they

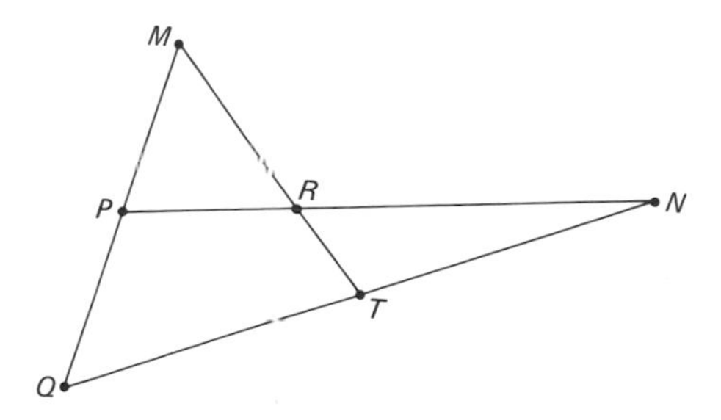

Figure 10.1

satisfy some previous condition of "alignment" and not in general. Apparently, then, the set S defined above might be different from $\mathscr{P}$. Indeed, we shall see that this can be the case. Replace Axiom P_4 with Pasch's Axiom. Assume that $\mathscr{P} - S \neq \varnothing$ and let $P' \in \mathscr{P} - S$. Then consider

$$S' = \{R \in \mathscr{P} : R \circ P' \vee Q \quad \text{for some} \quad Q \in S\}.$$

Now $S' = \mathscr{P}$ or $S' \neq \mathscr{P}$. In the latter case, let $P'' \in \mathscr{P} - S'$ and consider

$$S'' = \{R \in \mathscr{P} : R \circ P'' \vee Q \quad \text{for some} \quad Q \in S'\}.$$

Then $S'' = \mathscr{P}$ or $S'' \neq \mathscr{P}$. In the latter case, the next "step" is clear. Further, we have a reasonably good indication of how to take "successive steps" in the event the elements of $\mathscr{P}$ are not exhausted. In the theory of projective spaces, we shall be interested in a precise description of the process of taking these steps and a study of the structure of the resultant objects.

We shall find the following deviation from past conventions a convenience. For $(\mathscr{P}, \mathscr{L}, \circ)$ an incidence basis, we have been calling the elements of $\mathscr{P}$ "points" and the elements of $\mathscr{L}$ "lines." For the study of projective spaces, we should like to reserve these words for reference to certain subsets of $\mathscr{P}$. In particular, singleton sets $\{P\}$ for $P \in \mathscr{P}$ will be called "points" and sets of the form $\{P \in \mathscr{P} : P \circ l\}$ for $l \in \mathscr{L}$ will be called "lines" (compare with the elements of **P** and **L** in Exercise 3.14).

10.1. AXIOMS FOR A PROJECTIVE SPACE

An incidence basis $(\mathscr{P}, \mathscr{L}, \circ)$ will be called a *projective space* if the following axioms are satisfied.

AXIOM P_1. There exist $P \in \mathscr{P}$ and $l \in \mathscr{L}$ such that $P \not\circ l$.

AXIOM P_2. For $l \in \mathscr{L}$, there exist distinct $P, Q, R \in \mathscr{P}$ such that $P, Q, R \circ l$.

AXIOM P_3. If $P, Q \in \mathscr{P}$ such that $P \neq Q$, there exists one and only one $l \in \mathscr{L}$ such that $P, Q \circ l$.

For $P, Q, \in \mathscr{P}$ such that $P \neq Q$, we shall write, as before, $P \vee Q$ for the unique $l \in \mathscr{L}$ such that $P \circ l$, $Q \circ l$. In addition, $P \vee P = P$ for $P \in \mathscr{P}$.

AXIOM $P_4{}^*$. If $P, Q, R, T \in \mathscr{P}$ and distinct such that there exists $M \in P$ with $P \vee Q, R \vee T \circ M$, then there exists $N \in \mathscr{P}$ such that $P \vee R, Q \vee T \circ N$.

It is immediate that a projective plane is a projective space.

Throughout the remainder of this section let $(\mathscr{P}, \mathscr{L}, \circ)$ be a projective space.

Let $P \in \mathscr{P}$ and $S \subset \mathscr{P}$. Define for $S \neq \varnothing$,

$$\{P\} \vee' S = \{R \in \mathscr{P} : R \circ P \vee Q \quad \text{for some} \quad Q \in S\},$$

and $\{P\} \vee' \varnothing = \{P\}$. The singleton set $\{P\}$ for $P \in \mathscr{P}$ will be called a *point*; the set $\{P\} \vee' \{Q\}$ for P, Q, $\in \mathscr{P}$ such that $P \neq Q$ will be called a *line*; and the set $\{P\} \vee' L$ for $P \in \mathscr{P}$ and L a line such that $P \notin L$ will be called a *plane*. Further, for S, $T \subset \mathscr{P}$ such that each of S and T is a point, line, or plane, we shall write $S \circ T$ (read "S is on T") for $S \subset T$. Finally, for A a point and S a point, line or plane, we shall write $A \vee S$ for $A \vee' S$ and call $A \vee S$ the *join* of A and S.

We have the immediate corollary that for $P \in \mathscr{P}$ and $S \subset \mathscr{P}$, $S \subset \{P\} \vee' S$. Further, note that if L is a line, then $L = \{Q \in \mathscr{P} : Q \circ m\}$ for some $m \in \mathscr{L}$, and conversely.

THEOREM 10.1. If A, B are distinct points on a plane S, then $A \vee B \circ S$.

PROOF. Let $S = \{P\} \vee L$ be a plane, where $P \in \mathscr{P}$ and L is a line such that $\{P\} \not\circ L$. Let A', $B' \in \mathscr{P}$ such that $A' \neq B'$ and $\{A'\}$, $\{B'\} \circ \{P\} \vee L$. Now consider $P \neq A'$, B' and $\{A'\} \vee \{B'\} \neq L$; otherwise, the result is immediate. Finally, let $\{X\} \circ \{A'\} \vee \{B'\}$ for $X \in \mathscr{P}$ such that $X \neq A'$, B'. We have A'', $B'' \in L$ such that $A' \circ P \vee A''$ and $B' \circ P \vee B''$. In case $A' = A''$, we have $P \vee B''$, $X \vee A'' \circ B'$; so that by Axiom P_4*, $P \vee X$, $B'' \vee A'' \circ X'$ for some $X' \in \mathscr{P}$ (see Figure 10.2a). Hence $\{X\} \circ \{P\} \vee L$. In case $B' = B''$, the argument is similar. Now let $A' \neq A''$ and $B' \neq B''$ (see Figure 10.2b). Then $A' \vee B'$ and $A'' \vee B''$ are on some $C \in \mathscr{P}$, by virtue of Axiom P_4*, since $A' \vee A''$, $B' \vee B'' \circ P$. In case $X = C$, we have $X \circ P \vee C$, so that $\{X\} \circ \{P\} \vee L$. Let $X \neq C$. Then $P \vee X$, $A'' \vee C \circ X'$ for some $X' \in \mathscr{P}$ since $C \vee X$, $P \vee A'' \circ A'$. Hence $\{X\} \circ \{P\} \vee L$. Thus $\{A'\} \vee \{B'\} \circ S$.

THEOREM 10.2. If A is a point and L, M lines such that $A \not\circ L$ and $M \circ A \vee L$, then there exists a point B on L and M.

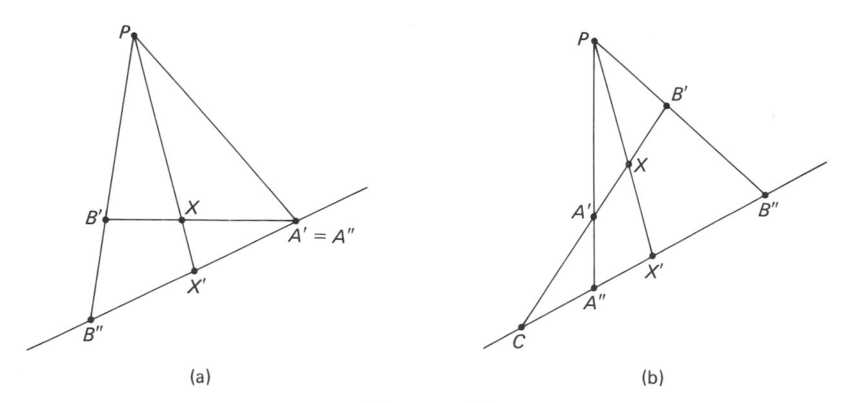

Figure 10.2

The proof is left as an exercise.

If distinct lines L, M are on a common point B, it readily follows from Axiom P_3 that B is unique. We shall call B the *intersection* of L and M and write $B = L \wedge M$.

THEOREM 10.3. If M, N are distinct lines on a plane, then M, N have a point of intersection.

PROOF. Let M, N be distinct lines on plane $\{P\} \vee L$, where $P \in \mathscr{P}$ and L is a line. Now let $M, N \neq L$; otherwise the previous theorem applies (see Figure 10.3). Now by the theorem, L intersects M in a point $\{Q\}$ and N in a point $\{R\}$. Let $Q' \in M$ such that $Q' \neq Q$ and $R' \in N$ such

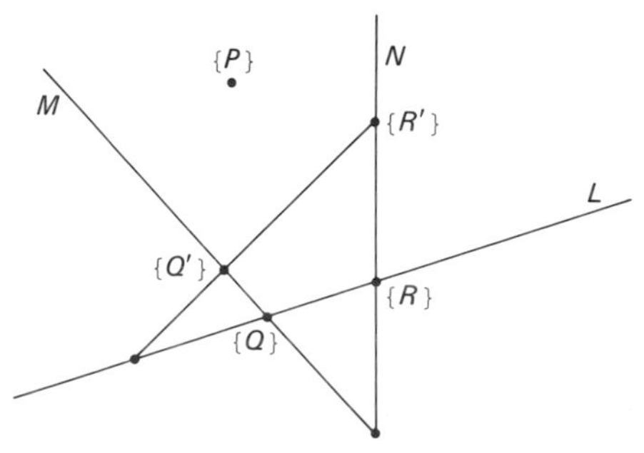

Figure 10.3

that $R' \neq R$. Further, let Q, Q', R, R' be distinct; otherwise, M, N have a point of intersection immediately. By the last theorem, $Q' \vee R'$ and $Q \vee R$ are on a common element of $\mathscr{P}$ since $\{Q\} \vee \{R\} = L$. Hence $Q \vee Q'$ and $R \vee R'$ are on a common element of $\mathscr{P}$. Thus $M = \{Q\} \vee \{Q'\}$ and $N = \{R\} \vee \{R'\}$ have a point of intersection.

COROLLARY 10.4. A plane in a projective space is a projective plane.

THEOREM 10.5. If a point A and a line L are on a plane S such that $A \not\circ L$, then $S = A \vee L$.

PROOF. Let A be a point and L a line on plane S such that $A \not\circ L$, and let $A = \{P\}$ for $P \in \mathscr{P}$. Now let $\{X\} \circ S$. In case $\{X\} = A$ or $\{X\} \circ L$, we have $\{X\} \circ A \vee L$. Let $\{X\} \neq A$ and $\{X\} \not\circ L$. Then $A \vee \{X\}$ is a line different from L. Define $\{X'\} = (A \vee \{X\}) \wedge L$. Then $A \neq \{X'\}$, $X \circ P \vee X'$ and $X' \in L$. Thus $\{X\} \circ A \vee L$. It is left as an exercise to show that an arbitrary point on $A \vee L$ is also on S. We shall consider the proof complete.

COROLLARY 10.6. Distinct noncollinear points A, B, C are on one and only one plane.

We shall write $A \vee B \vee C$ for this plane. The proof of the corrollary is left as an exercise.

THEOREM 10.7. If M is a line on distinct planes S and T, then any point on S and T is on M.

COROLLARY 10.8. If M is a line on distinct planes S and T, then M is the only line on S and T.

The line M of the corollary will be called the *intersection* of S and T and we shall write $M = S \wedge T$. The proofs of the theorem and corollary are left as exercises.

A triangle is defined as before. The perspectivity of two triangles from a point or a line has the previous meaning. The statement of the Theorem of Desargues is unchanged. Now we are in a position to state the following interesting result.

THEOREM 10.9. The Theorem of Desargues holds in any projective space which is not a projective plane.

PROOF. Let triangles ABC and $A'B'C'$ be perspective from point O in a projective space which is not a projective plane. Define $A_1 = (B \vee C) \wedge (B' \vee C')$ (the intersection exists since $B \vee C$ and $B' \vee C'$ are on plane $O \vee B \vee C$), $B_1 = (C \vee A) \wedge (C' \vee A')$, and $C_1 = (A \vee B) \wedge (A' \vee B')$. In addition, define $S = A \vee B \vee C$ and $S' = A' \vee B' \vee C'$. We shall consider two cases.

CASE 1. Let $S \neq S'$ (see Figure 10.4). Then A_1, B_1, C_1 are distinct

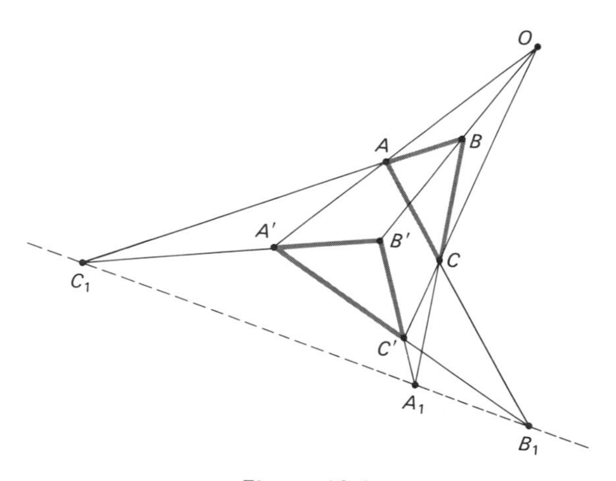

Figure 10.4

and on both S and S'. Hence $S \wedge S' = A_1 \vee B_1$ and $C_1 \circ A_1 \vee B_1$. Thus ABC and $A'B'C'$ are perspective from $S \wedge S'$.

CASE 2. Let $S = S'$ (see Figure 10.5). Then $O \circ S$. Now there exist points O_1 and O_2 such that $O_1 \not\circ S$ and $O_2 \circ O \vee O_1$ with $O_2 \neq O, O_1$. Further, $O_1 \vee A$ and $O_2 \vee A'$ are distinct and on $O \vee O_1 \vee A$, so

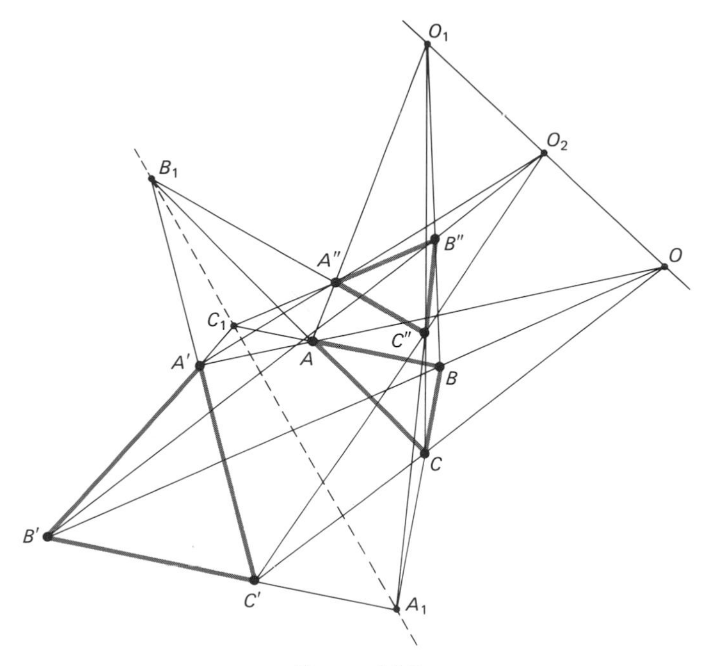

Figure 10.5

that $A'' = (O_1 \vee A) \wedge (O_2 \vee A')$ exists as a point. Similarly, $B'' = (O_1 \vee B) \wedge (O_2 \vee B')$ and $C'' = (O_1 \vee C) \wedge (O_2 \vee C')$ exist. Further, A'', B'', C'' are distinct and noncollinear. Now triangles ABC and $A''B''C''$ are perspective from O_1, so that by Case 1, they are perspective from $M = S \wedge (A'' \vee B'' \vee C'')$. Similarly, triangles $A'B'C'$ and $A''B''C''$ are perspective from M. Finally, C_1 is on planes $O_1 \vee A \vee B$ and $O_2 \vee A' \vee B'$. In addition, $A'' \vee B''$ is on these planes. Hence $C_1 \circ A'' \vee B''$. Similarly, $B_1 \circ A'' \vee C''$ and $A_1 \circ B'' \vee C''$. Thus $A_1, B_1, C_1 \circ M$, that is, triangles ABC and $A'B'C'$ are perspective from M.

EXERCISES

10.1. Show that for $P \in \mathscr{P}$ and $S \subset \mathscr{P}$, $S \subset \{P\} \vee' S$.

10.2. Show that if L is a line, then $L = \{Q \in P : Q \circ m\}$ for some $m \in \mathscr{L}$, and conversely.

10.3. Prove Theorem 10.2.

10.4. Prove Corollary 10.4.

10.5. Complete the proof of Theorem 10.5.

10.6. Prove Corollary 10.6.

10.7. Prove Theorem 10.7.

10.8. Prove Corollary 10.8.

10.9. In the proof of Theorem 10.9, show that A'', B'', C'' are distinct and noncollinear.

10.10. Show that any two planes in a projective space are isomorphic. (*Hint*: Consider, first, the case when the planes have a line of intersection.)

10.2. EXAMPLES

We shall consider two algebraic examples of projective spaces. Throughout this section let $(\mathscr{F}, +, \cdot)$ be a field.

BASIS 10.1. Let $n \geqq 3$. Define $\mathscr{P}$ to be the set of all n-tuples $(x_1, \ldots, x_n)$ in $\mathscr{F}$ such that not all x_i are equal to 0. Now identify $(x_1, \ldots, x_n)$, $(y_1, \ldots, y_n) \in \mathscr{P}$, written $(x_1, \ldots, x_n) \equiv (y_1, \ldots, y_n)$, if $x_i = y_i k$ for some nonzero $k \in \mathscr{F}$ and $i = 1, \ldots, n$. A "line" will be a set of the form

$$\{(x_1 j + y_1 k, \ldots, x_n j + y_n k) : j, k \in \mathscr{F} \text{ such that not both } j = 0, k = 0\}$$

where $(x_1, \ldots, x_n), (y_1, \ldots, y_n) \in \mathscr{P}$ such that $(x_1, \ldots, x_n) \not\equiv (y_1, \ldots, y_n)$. Then $\mathscr{L}$ is the set of all such lines. Finally, for $P \in \mathscr{P}$ and $l \in \mathscr{L}$, $P \circ l$ will mean $P \in l$.

It will not be verified directly that $(\mathscr{P}, \mathscr{L}, \circ)$ is a projective space. Actually, this incidence basis is a special case of the next basis to be considered. The details will be supplied in the verification that the latter is a projective space.

BASIS 10.2. Let S be any set containing at least three elements. Define $\mathscr{P}$ to be the set of all functions on S to $\mathscr{F}$ excluding the function with constant value 0.

For $F, G \in \mathscr{P}$ and $k \in \mathscr{F}$, define functions $F + G$ and $F \cdot k$ by

$$(F + G)(x) = F(x) + G(x) \qquad \text{for} \quad x \in S,$$
$$(F \cdot k)(x) = F(x) \cdot k \qquad \text{for} \quad x \in S.$$

Now identify $F, G \in \mathscr{P}$, written $F \equiv G$, if $F = G \cdot k$ for some $k \in \mathscr{F}$ such that $k \neq 0$. Further, for $F, G \in \mathscr{P}$, define

$$F|G = \{F \cdot j + G \cdot k : j, k \in \mathscr{F} \quad \text{such that not both} \quad j = 0, k = 0\}.$$

Then for $F, G, F', G' \in \mathscr{P}$, $F|G = F'|G'$ if $F \equiv F'$ and $G \equiv G'$. Now define

$$\mathscr{L} = \{F|G : F, G \in \mathscr{P} \quad \text{such that} \quad F \not\equiv G\}.$$

Then for $P, P' \in \mathscr{P}$ such that $P \equiv P'$ and $l \in \mathscr{L}$, $P \in l$ if and only if $P' \in l$. Finally, define for $P \in \mathscr{P}$ and $l \in \mathscr{L}$, $P \circ l$ to mean $P \in l$. Then the definition of the elements of $\mathscr{L}$ and of incidence are independent of the identification of points in $\mathscr{P}$. Thus $(\mathscr{P}, \mathscr{L}, \circ)$ is a well-defined incidence basis.

For Axiom P_1, let x_1, x_2, x_3 be distinct elements in S and define F_i for $i = 1, 2, 3$ by the following: For $x \in S$,

$$F_i(x) = \begin{cases} 1 & \text{if} \quad x = x_i, \\ 0 & \text{if} \quad x \neq x_i. \end{cases}$$

Then $F_1 \not\equiv F_2$ and $F_3 \not\circ F_1|F_2$.

For Axiom P_2, let $l \in \mathscr{L}$, that is, $l = F|G$ for some $F, G \in \mathscr{P}$ such that $F \not\equiv G$. Then $F, G, F + G \circ l$ and are distinct.

For Axiom P_3, let $F, G \in \mathscr{P}$ such that $F \not\equiv G$. Then $F, G \circ F|G$. To show the uniqueness of $F|G$, let $F', G' \in \mathscr{P}$ such that $F' \not\equiv G'$ and $F, G \circ F'|G'$. Then

$$F = F' \cdot j + G' \cdot k \qquad \text{for some} \quad j, k \in \mathscr{F} \text{ such that } (j, k) \neq (0, 0)$$

and

$$G = F' \cdot m + G' \cdot n \qquad \text{for some} \quad m, n \in \mathscr{F} \quad \text{such that } (m, n) \neq (0, 0).$$

Let $P \circ F|G$. Then $P = F \cdot p + G \cdot q$ for some p, q not both 0. Now $P = (F' \cdot j + G' \cdot k) \cdot p + (F' \cdot m + G' \cdot n) \cdot q = F' \cdot (jp + mq) + G' \cdot (kp + nq)$. Suppose $jp + mq = 0$ and $kp + nq = 0$. Then $jp = -mq$ and $kp = -nq$. In case $j \neq 0$, $p = -j^{-1}mq$, so that $k(-j^{-1}mq) = -nq$.

Hence $n = kj^{-1}m$. Now $G = F' \cdot m + G' \cdot n = F' \cdot m + G' \cdot (kj^{-1}m) \equiv F' \cdot j + G' \cdot k = F$, a contradiction. In like manner, we arrive at a contradiction in case $k \neq 0$. Hence not both $jp + mq = 0$ and $kp + nq = 0$ hold, so that $P \circ F'|G'$. Thus $F|G \subset F'|G'$. To show that F', $G' \circ F|G$, consider the case when $j \neq 0$. Then $F \cdot j^{-1} = F' + G' \cdot (kj^{-1})$, so that $F' = F \cdot j^{-1} + G' \cdot (-kj^{-1})$. We also have

$$\begin{aligned} G &= (F \cdot j^{-1} + G' \cdot (-kj^{-1})) \cdot m + G' \cdot n \\ &= F \cdot (j^{-1}m) + G' \cdot (n - kj^{-1}m). \end{aligned}$$

Now $n - kj^{-1}m \neq 0$; otherwise, we have from the above argument that $G \equiv F$, a contradiction. Hence

$$G' = F \cdot ((-j^{-1}m)(n - kj^{-1}m)^{-1}) + G \cdot (n - kj^{-1}m)^{-1}$$

with $n - kj^{-1}m \neq 0$, and

$$\begin{aligned} F' &= F \cdot j^{-1} + (F \cdot ((-j^{-1}m)(n - kj^{-1}m)^{-1}) \\ &\qquad\qquad + G \cdot (n - kj^{-1}m)^{-1}) \cdot (-kj^{-1}) \\ &= F \cdot (j^{-1} + (j^{-1}m)(n - kj^{-1}m)^{-1}(kj^{-1})) \\ &\qquad\qquad + G \cdot (-(n - kj^{-1}m)^{-1}(kj^{-1})). \end{aligned}$$

Further, if $-(n - kj^{-1}m)^{-1}(kj^{-1}) = 0$, we have $k = 0$, so that

$$j^{-1} + (j^{-1}m)(n - kj^{-1}m)^{-1}(kj^{-1}) = j^{-1} \neq 0.$$

Hence F', $G' \circ F|G$. Similarly, we obtain F', $G' \circ F|G$ in case $k \neq 0$. Now by the first half of the argument we have the reverse inclusion $F'|G' \subset F|G$. Thus $F'|G' = F|G$. Now we can write $F \vee G = F|G$ for $F \not\equiv G$.

For Axiom P_4*, let $F, G, H, I \in \mathscr{P}$ and be distinct such that $P \circ F \vee G$, $H \vee I$ for some $P \in \mathscr{P}$. It is left as an exercise to show the existence of $Q \in \mathscr{P}$ such that $Q \circ F \vee H$, $G \vee I$.

We have shown that $(\mathscr{P}, \mathscr{L}, \circ)$ of Basis 10.2 is a projective space. This includes Basis 10.1 if we let $S = \{1, \ldots, n\}$.

EXERCISES

10.11. In the verification that Basis 10.2 is a projective space, show the following.

(a) For F, $G \in \mathscr{P}$ and m, $n \in \mathscr{F}$, $(F \cdot m) \cdot n = F \cdot (mn)$, $F \cdot m + F \cdot n = F \cdot (m + n)$, and $F \cdot m + G \cdot m = (F + G) \cdot m$.

(b) For F, G, F', $G' \in \mathscr{P}$ such that $F \equiv F'$ and $G \equiv G'$, $F|G = F'|G'$.

(c) For P, $P' \in \mathscr{P}$ such that $P \equiv P'$ and $l \in \mathscr{L}$, $P \in l$ if and only if $P' \in 1$.

(d) For Axiom P_1, show that $F_1 \not\equiv F_2$ and $F_3 \not\in F_1|F_2$.

(e) For Axiom P_2, show that F, G, $F + G$ are distinct.

(f) Complete the verification of Axiom P_4*.

10.12. What connection can be established between Basis 3.5 and Basis 10.1 with $n = 3$?

PROJECTIVE SPACES—PART 2 CHAPTER 11

We shall continue the study of projective spaces with the consideration of "higher level" sets formed under our join operation $\vee'$. Some of the new results will resemble the results of the last chapter and, of course, will represent generalizations of these earlier results. Throughout this chapter we shall assume that $(\mathscr{P}, \mathscr{L}, \circ)$ is a projective space and adopt the notation of the last chapter.

11.1. SUBSPACES AND DIMENSION

LEMMA 11.1. For $P, Q \in \mathscr{P}$ and $S \subset \mathscr{P}$,

$$\{P\} \vee' (\{Q\} \vee' S) = \{Q\} \vee' (\{P\} \vee' S).$$

PROOF. Let $P, Q \in \mathscr{P}$ and $S \subset \mathscr{P}$. Consider the case when $P \neq Q$. Let $X \in \{P\} \vee' (\{Q\} \vee' S)$. Then there exists $R \in \{Q\} \vee' S$ such that $X \circ P \vee R$. Now there exists $T \in S$ such that $R \circ Q \vee T$ (see Figure 11.1). In case $\{Q\} \mathbf{v} \{X\}, \{P\} \mathbf{v} \{T\}$ are lines and $\{P\} \mathbf{v} \{Q\} \mathbf{v} \{T\}$ is a plane (the remaining cases are left as an exercise), the lines intersect in a point $\{Y\}$ for some $Y \in \mathscr{P}$. Hence $Y \circ P \vee T$ and $X \circ Q \vee Y$. Now

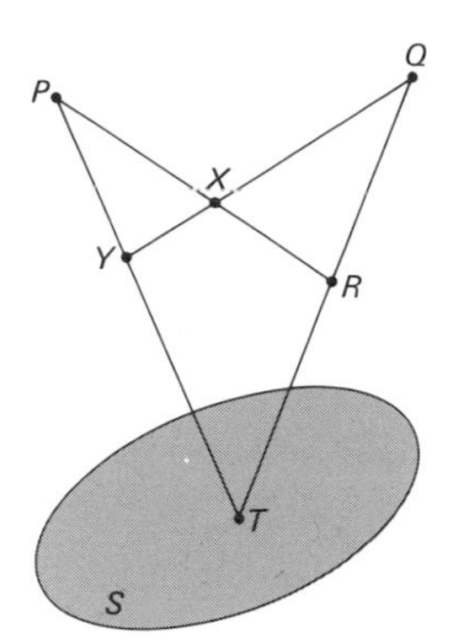

Figure 11.1

$Y \in \{P\} \vee' S$ and $X \in \{Q\} \vee' (\{P\} \vee' S)$. Thus $\{P\} \vee' (\{Q\} \vee' S) \subset \{Q\} \vee' (\{P\} \vee' S)$. The reverse inclusion follows similarly. The proof is complete.

Let (A_n) be a sequence of points. Inductively define the sequence $(\bigvee_{i=1}^{n} A_i)$ of subsets of $\mathscr{P}$ by the following:

(a) $\bigvee_{i=1}^{1} A_i = A_1$;
(b) for n a positive integer, $\bigvee_{i=1}^{n+1} A_i = A_{n+1} \vee' \bigvee_{i=1}^{n} A_i$.

The next theorem states that $\bigvee_{i=1}^{n} A_i$ does not depend on the order of the A_i's.

THEOREM 11.2. If $\{A_1, \ldots, A_n\}$ and $\{B_1, \ldots, B_n\}$ are equal sets of points, then $\bigvee_{i=1}^{n} A_i = \bigvee_{i=1}^{n} B_i$.

PROOF. The proof will be by induction. We have $A_1 = B_1$ if $\{A_1\} = \{B_1\}$. Hence $\bigvee_{i=1}^{1} A_i = A_1 = B_1 = \bigvee_{i=1}^{1} B_i$, and thus the statement holds for $n = 1$. Let the statement hold for $n = k$ and let $\{A_1, \ldots, A_{k+1}\} = \{B_1, \ldots, B_{k+1}\}$. Then for some j such that $1 \leqq j \leqq k+1$, $B_{k+1} = A_j$. In case $j = k+1$, we have

$$\bigvee_{i=1}^{k+1} A_i = A_{k+1} \vee' \bigvee_{i=1}^{k} A_i = B_{k+1} \vee' \bigvee_{i=1}^{k} B_i = \bigvee_{i=1}^{k+1} B_i$$

where $\bigvee_{i=1}^{k} A_i = \bigvee_{i=1}^{k} B_i$ by the induction hypothesis since $\{A_1, \ldots, A_k\} = \{B_1, \ldots, B_k\}$. Now consider $j \neq k+1$. Define

$$C_i = \begin{cases} A_i & \text{for } 1 \leqq i < j, \\ A_{i+1} & \text{for } j \leqq i < k, \\ A_j & \text{for } i = k \end{cases}$$

and

$$D_i = \begin{cases} A_i & \text{for } 1 \leqq i < j, \\ A_{i+1} & \text{for } j \leqq i \leqq k. \end{cases}$$

Then $C_i = D_i$ for $1 \leqq i < k$. Further $\{C_1, \ldots, C_k\} = \{A_1, \ldots, A_k\}$ and $\{D_1, \ldots, D_k\} = \{B_1, \ldots, B_k\}$. Hence by the induction hypothesis, $\bigvee_{i=1}^{k} C_i = \bigvee_{i=1}^{k} A_i$ and $\bigvee_{i=1}^{k} D_i = \bigvee_{i=1}^{k} B_i$. Now

$$\begin{aligned}
\bigvee_{i=1}^{k+1} A_i &= A_{k+1} \vee' \bigvee_{i=1}^{k} A_i \\
&= A_{k+1} \vee' \bigvee_{i=1}^{k} C_i \\
&= A_{k+1} \vee' (C_k \vee' \bigvee_{i=1}^{k-1} C_i) \\
&= A_{k+1} \vee' (A_j \vee' \bigvee_{i=1}^{k-1} C_i) \\
&= A_j \vee' (A_{k+1} \vee' \bigvee_{i=1}^{k-1} D_i) \\
&= A_j \vee' (D_k \vee' \bigvee_{i=1}^{k-1} D_i) \\
&= A_j \vee' \bigvee_{i=1}^{k} D_i \\
&= B_{k+1} \vee' \bigvee_{i=1}^{k} B_i \\
&= \bigvee_{i=1}^{k+1} B_i .
\end{aligned}$$

Thus the statement holds for $n = k + 1$. This completes the proof.

For $A_1, \ldots, A_n$ points, we define $\langle A_1, \ldots, A_n \rangle = \bigvee_{i=1}^{n} A_i$, called the *join* of $A_1, \ldots, A_n$. We have as an immediate corollary $\langle A_1, \ldots, A_n \rangle = \langle B_1, \ldots, B_n \rangle$ if $\{A_1, \ldots, A_n\} = \{B_1, \ldots, B_n\}$.

Let S be a set of points and $T \subset \mathscr{P}$. We shall say that S *spans* T (or T is *spanned by* S) if

(a) $\langle A_1, \ldots, A_n \rangle \subset T$ for any points $A_1, \ldots, A_n \in S$;
(b) for any point $A \subset T$, there exist points $B_1, \ldots, B_m \in S$ such that $A \subset \langle B_1, \ldots, B_m \rangle$.

A set of points S spans a unique subset T of $\mathscr{P}$, where

$$T = \{P \in \mathscr{P} : \{P\} \subset \langle A_1, \ldots, A_n \rangle \quad \text{for some} \quad A_1, \ldots, A_n \in S\}.$$

We have $\langle A_1, \ldots, A_n \rangle$ spanned by $\{A_1, \ldots, A_n\}$, $\varnothing$ spanned by $\varnothing$, $\{A\}$ spanned by $\{A\}$ for A a point, and $\mathscr{P}$ spanned by $\{\{P\} : P \in \mathscr{P}\}$. Further, if S spans T, then $A \in S$ implies $A \subset T$.

A set spanned by a finite set of points will be called a *subspace*. We include $\varnothing$ among our subspaces. Now define $\overline{\mathscr{P}}$ to be the set of all subspaces contained in $\mathscr{P}$. If $S, T \in \overline{\mathscr{P}}$ such that $S \subset T$, we shall write $S \circ T$ (read "S is on T"). In this section, we extend our join $\vee$ and intersection $\wedge$ to any pair of elements of $\overline{\mathscr{P}}$. The study of our projective space $(\mathscr{P}, \mathscr{L}, \circ)$ will become, then, a study of the structure of $\overline{\mathscr{P}}$ relative to $\circ$, $\vee$, and $\wedge$.

Any point, line, or plane is a subspace and the new meaning of $\circ$ is in agreement with the meaning ascribed to that symbol in the last chapter. In addition, if $\{A_1, \ldots, A_m\}$ is a subset of the set $\{B_1, \ldots, B_n\}$ of points, then $\langle A_1, \ldots, A_m \rangle \circ \langle B_1, \ldots, B_n \rangle$.

LEMMA 11.3. If $A_1, \ldots, A_n, B$ are points such that $B \circ \langle A_1, \ldots, A_n \rangle$, then $B \circ A_n \vee C$ for some point $C \circ \langle A_1, \ldots, A_{n-1} \rangle$.

The proof is left as an exercise.

LEMMA 11.4. If $A_1, \ldots, A_n$, B, C are points such that $B, C \circ \langle A_1, \ldots, A_n \rangle$, then $B \vee C \circ \langle A_1, \ldots, A_n \rangle$.

PROOF. The proof will be inductive. The statement obviously holds for $n = 1$. Let the statement hold for $n = k$ and let $B, C \circ \langle A_1, \ldots, A_{k+1} \rangle$. Then by Lemma 11.3, there exist points B' and C' on $\langle A_1, \ldots, A_k \rangle$ such that $B \circ A_{k+1} \vee B'$ and $C \circ A_{k+1} \vee C'$ (see Figure 11.2). Consider the case when A_{k+1}, B, C are distinct and

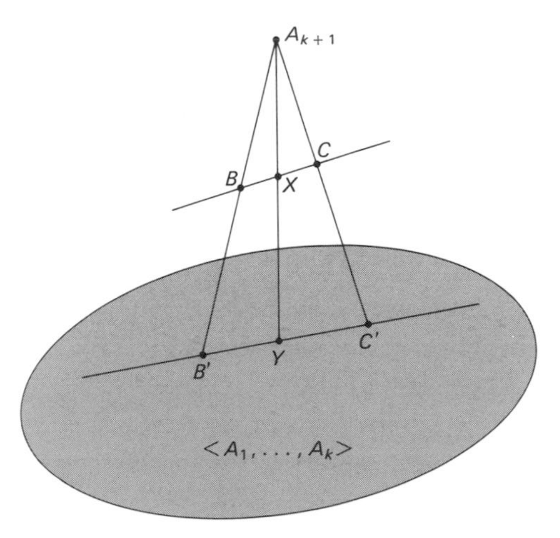

Figure 11.2

noncollinear. (The remaining cases are left as an exercise.) Then $B' \neq C'$ and by the induction hypothesis, $B' \vee C' \circ \langle A_1, \ldots, A_k \rangle$. Now let $X \circ B \vee C$. Then lines $A_{k+1} \vee X$ and $B' \vee C'$ are on plane $A_{k+1} \vee B' \vee C'$ and therefore have a point of intersection Y, where $Y \circ \langle A_1, \ldots, A_k \rangle$. Now $X \circ A_{k+1} \vee Y$; hence $X \circ \langle A_1, \ldots, A_{k+1} \rangle$. Thus $B \vee C \circ \langle A_1, \ldots, A_{k+1} \rangle$. This completes the proof.

THEOREM 11.5. If $A_1, \ldots, A_n$, A are points such that $A \circ \langle A_1, \ldots, A_n \rangle$ and $A \not\circ \langle A_1, \ldots, A_{n-1} \rangle$, then $\langle A_1, \ldots, A_n \rangle = \langle A_1, \ldots, A_{n-1}, A \rangle$.

PROOF. Let $A \circ \langle A_1, \ldots, A_n \rangle$ such that $A \not\circ \langle A_1, \ldots, A_{n-1} \rangle$. Now let $A \neq A_n$. Then by Lemma 11.3, there exists $B \circ \langle A_1, \ldots, A_{n-1} \rangle$ such that $A \circ A_n \vee B$. Further, $A \neq B$ since $A \not\circ \langle A_1, \ldots, A_{n-1} \rangle$ (see Figure 11.3). Now let X be a point on $\langle A_1, \ldots, A_n \rangle$. Then there exists $C \circ \langle A_1, \ldots, A_{n-1} \rangle$ such that $X \circ A_n \vee C$. Consider the case when $X \not\circ A_n \vee B$ and $X \neq A_n, C$. (The remaining cases are left as an exercise.) Then lines $A \vee X$ and $B \vee C$ are on plane $A_n \vee B \vee C$ and therefore intersect in a point D. By Lemma 11.4, $D \circ \langle A_1, \ldots, A_{n-1} \rangle$. Hence

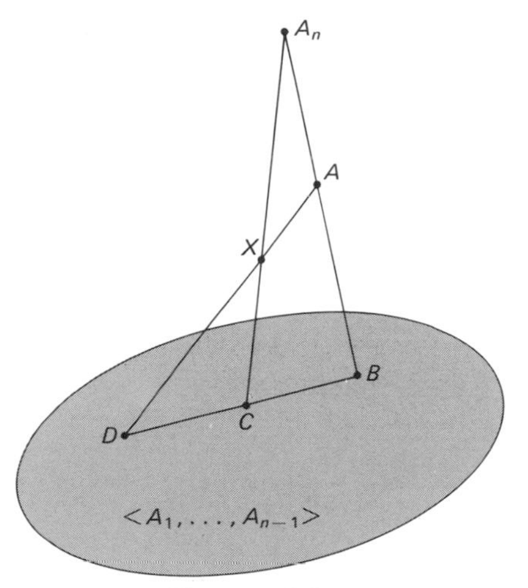

Figure 11.3

$X \circ A \vee' \langle A_1, \ldots, A_{n-1} \rangle = \langle A_1, \ldots, A_{n-1}, A \rangle$. Thus $\langle A_1, \ldots, A_n \rangle \subset \langle A_1, \ldots, A_{n-1}, A \rangle$. The reverse inclusion is left as an exercise. This completes the proof.

THEOREM 11.6. If $A_1, \ldots, A_n$ are points on a subspace S, then $\langle A_1, \ldots, A_n \rangle \circ S$.

PROOF. We shall proceed inductively. Let S be a subspace. If $A_1 \circ S$, we have $\langle A_1 \rangle = A_1 \circ S$. Hence the statement holds for $n = 1$. Let the statement hold for $n = k$, and let $A_1, \ldots, A_{k+1} \circ S$. Then by the induction hypothesis, $\langle A_1, \ldots, A_k \rangle \circ S$. We want to show $\langle A_1, \ldots, A_{k+1} \rangle \circ S$. The completion of the proof is left as an exercise.

THEOREM 11.7. Let $A_1, \ldots, A_n$ be points. If $A_n \circ \langle B_1, \ldots, B_m \rangle$ for $\{B_1, \ldots, B_m\} \subset \{A_1, \ldots, A_{n-1}\}$, then $\langle A_1, \ldots, A_n \rangle = \langle A_1, \ldots, A_{n-1} \rangle$.

The proof is left as an exercise.

Let us consider repeated applications of the last theorem, as long as the hypothesis is satisfied, to the subspace spanned by a set of points in order to reduce the number of elements in the spanning set, in particular, to obtain a minimal spanning set. Our next concept facilitates the description of this situation.

A set $\{A_1, \ldots, A_n\}$ for which $A_i \neq A_j$ for $i \neq j$ is said to be *independent* if $A_i \not\!o \langle A_1, \ldots, A_{i-1}, A_{i+1}, \ldots, A_n \rangle$ for $i = 1, \ldots, n$. The empty set is independent. In addition, if A, B are points, then $\{A\}$ and $\{A, B\}$ are independent. Further, any subset of an independent set is independent.

A set of points is said to be *dependent* if it is not independent. Let $S = \{A_1, \ldots, A_n\}$ be a set of points. The condition $A_i \circ \langle A_1, \ldots, A_{i-1}, A_{i+1}, \ldots, A_n \rangle$ for some i is necessary for S to be dependent, but it is not sufficient. (Why?)

A set of points which is independent and spans a subspace S will be called a *basis* of S.

LEMMA 11.8. If $A_1, \ldots, A_{n+1}$ are distinct points on $\langle B_1, \ldots, B_n \rangle$, then $\{A_1, \ldots, A_{n+1}\}$ is dependent.

PROOF. The proof will be by induction. The statement is vacuously true for $n = 1$. Let the statement hold for $n = k$ and let $A_1, \ldots, A_{k+2}$ be distinct points on $S = \langle B_1, \ldots, B_{k+1} \rangle$. In case $A_1, \ldots, A_{k+1} \circ \langle B_2, \ldots, B_{k+1} \rangle$, we have, by the induction hypothesis, $\{A_1, \ldots, A_{k+1}\}$ dependent. Hence $\{A_1, \ldots, A_{k+2}\}$ is dependent. In case $A_1, \ldots, A_{k+1} \circ \langle B_2, \ldots, B_{k+1} \rangle$ does not hold, we can assume without loss of generality that $A_1 \not\!o \langle B_2, \ldots, B_{k+1} \rangle$. Then by Theorem 11.5, $S = \langle A_1, B_2, \ldots, B_{k+1} \rangle$. In case $A_1, \ldots, A_{k+1} \circ \langle A_1, B_3, \ldots, B_{k+1} \rangle$, we have, as before, $\{A_1, \ldots, A_{k+2}\}$ dependent. Otherwise, we can assume without loss of generality that $A_2 \not\!o \langle A_1, B_3, \ldots, B_{k+1} \rangle$ and we have, as before, $S = \langle A_1, A_2, B_3, \ldots, B_{k+1} \rangle$. Proceeding in this manner, we obtain

(a) for some j with $1 \leqq j \leqq k+1$, $S = \langle A_1, \ldots, A_{j-1}, B_j, \ldots, B_{k+1} \rangle$ and $A_1, \ldots, A_{k+1} \circ \langle A_1, \ldots, A_{j-1}, B_{j+1}, \ldots, B_{k+1} \rangle$ or

(b) $S = \langle A_1, \ldots, A_k, B_{k+1} \rangle$ and $A_{k+1} \not\!o \langle A_1, \ldots, A_k \rangle$.

If (a) holds, we have $\{A_1, \ldots, A_{k+1}\}$ dependent. Hence $\{A_1, \ldots, A_{k+2}\}$ dependent. If (b) holds, we have $S = \langle A_1, \ldots, A_{k+1} \rangle$. Hence $\{A_1, \ldots, A_{k+2}\}$ is dependent since $A_{k+2} \circ S$. Thus the statement holds for $n = k + 1$. This completes the proof.

THEOREM 11.9. If $A_1, \ldots, A_n$ are distinct points on a subspace such that $\{A_1, \ldots, A_n\}$ is independent, then there is a basis of the subspace containing $\{A_1, \ldots, A_n\}$.

An immediate corollary is that every nonempty subspace will have a basis since $\{A\}$ is independent for A a point on the subspace.

PROOF OF THEOREM 11.9. Let $\{A_1, \ldots, A_n\}$ be an independent set of points on a subspace S such that $A_i \neq A_j$ for $i \neq j$. By definition, $S = \langle B_1, \ldots, B_m \rangle$ for some points $B_1, \ldots, B_m$. Consider all independent sets of points of the form $\{A_1, \ldots, A_n, C_1, \ldots, C_p\}$ with $C_1, \ldots, C_p \circ S$, $C_i \neq C_j$ for $i \neq j$ and $A_i \neq C_j$. By Lemma 11.8, $n + p \leqq m$. Hence among these sets there will be one with a maximum p. Let $T = \{A_1, \ldots, A_n, A_{n+1}, \ldots, A_k\}$ be such a set. Then $n \leqq k \leqq m$. To show that T spans S, let $A \circ S$. Suppose $A \not\circ \langle A_1, \ldots, A_k \rangle$. Then, also, $A_j \not\circ \langle A_1, \ldots, A_{j-1}, A, A_{j+1}, \ldots, A_k \rangle$ for $j = 1, \ldots, k$. Otherwise, if $A_j \circ \langle A_1, \ldots, A_{j-1}, A, A_{j+1}, \ldots, A_k \rangle$ for some j, then by Theorem 11.5, $\langle A_1, \ldots, A_{j-1}, A, A_{j+1}, \ldots, A_k \rangle = \langle A_1, \ldots, A_k \rangle$ since $A_j \not\circ \langle A_1, \ldots, A_{j-1}, A_{j+1}, \ldots, A_k \rangle$. Hence $A \circ \langle A_1, \ldots, A_k \rangle$, a contradiction. Therefore $\{A_1, \ldots, A_k, A\}$ is independent contrary to the definition of k. Thus $A \circ \langle A_1, \ldots, A_k \rangle$ and $S \subset \langle A_1, \ldots, A_k \rangle$. The reverse inclusion is immediate. Thus T spans S.

THEOREM 11.10. A set $S \subset \mathscr{P}$ is a subspace if and only if

(a) there exist points $B_1, \ldots, B_m$ such that $S \subset \langle B_1, \ldots, B_m \rangle$;
(b) $A \vee B \subset S$ for points $A, B \subset S$.

PROOF. The "only if" part is left as an exercise. For the converse, let $S \subset \langle B_1, \ldots, B_m \rangle$ and $A \vee B \subset S$ for points $A, B \subset S$. If $S = \varnothing$, we are finished. Let $S \neq \varnothing$. Then there is at least one set of independent points in S, for example, $\{A\}$ for A a point in S. Further, if $\{A_1, \ldots, A_n\}$ is an independent set of points in S with the A_i distinct, then $n \leqq m$ by Lemma 11.8. Hence there is a maximal such set $\{A_1, \ldots, A_k\}$. Following the proof of the last theorem, we have $S \subset \langle A_1, \ldots, A_k \rangle$. The reverse inclusion is left for the exercise. This completes the proof.

THEOREM 11.11. If $\{A_1, \ldots, A_m\}$ and $\{B_1, \ldots, B_n\}$ are bases of a subspace such that $A_i \neq A_j$ and $B_i \neq B_j$ for $i \neq j$, then $m = n$.

PROOF. Let $\{A_1, \ldots, A_m\}$ and $\{B_1, \ldots, B_n\}$ be bases of a subspace S with $A_i \neq A_j$ and $B_i \neq B_j$ for $i \neq j$. Suppose $n < m$. Then $n + 1 \leqq m$. Hence by Lemma 11.8, $\{A_1, \ldots, A_{n+1}\}$ is dependent contrary to $\{A_1, \ldots, A_m\}$ being independent. Thus $n \geqq m$. Similarly, we can show that $m \geqq n$. Thus $m = n$.

We shall define a subspace S with a basis $\{A_1, \ldots, A_n\}$ such that $A_i \neq A_j$ for $i \neq j$ to have *dimension* $n - 1$ and write $\dim S = n - 1$. Thus $\varnothing$, points, lines, and planes have dimension $-1, 0, 1$, and 2, respectively. The projective space $(\mathscr{P}, \mathscr{L}, \circ)$ is said to be *n-dimensional* if $\mathscr{P}$ is a subspace having dimension n.

COROLLARY 11.12.

(a) If $(\mathscr{P}, \mathscr{L}, \circ)$ is n-dimensional and S is a subspace, then $S \neq \mathscr{P}$ if and only if $\dim S < n$.
(b) If $S, T \in \overline{\mathscr{P}}$ such that $S \circ T$ and $\dim S = \dim T$, then $S = T$.
(c) The elements of a subspace having dimension $k \geqq 2$ form a k-dimensional projective space.

In part (c) of the corollary, the elements of a subspace S are interpreted to be an incidence basis in the following sense. Define $\mathbf{P} = \{A \in \overline{\mathscr{P}} \colon A \circ S \text{ and } \dim A = 0\}$, $\mathbf{L} = \{L \in \overline{\mathscr{P}} \colon L \circ S \text{ and } \dim L = 1\}$, and for $B \in \mathbf{P}$ and $M \in \mathbf{L}$, $B \odot M$ to mean $B \circ M$. Then $(\mathbf{P}, \mathbf{L}, \odot)$ is the incidence basis being considered in the corollary. The proof of the corollary is left as an exercise.

THEOREM 11.13. The set-theoretic intersection of two subspaces is a subspace.

PROOF. Let S' and S'' be subspaces and $S = S' \cap S''$. Now let $\{A_1, \ldots, A_n\}$ be a basis of S'. Then $S \subset \langle A_1, \ldots, A_n \rangle$. In case $S = \varnothing$,

S is a subspace. Let $S \neq \varnothing$ and let $B, C \subset S$. Then $B, C \circ S'$ and $B, C \circ S''$. Therefore by Lemma 11.4, $B \vee C \circ S', S''$. Hence $B \vee C \subset S$. Thus by Theorem 11.10, S is a subspace.

For S and T subspaces, we shall call $S \cap T$ the *intersection* (in the geometric sense) *of* S *and* T and write $S \wedge T$ for $S \cap T$. The present definition of intersection is in agreement with that made in the last chapter for subspaces that are lines or planes. Two subspaces S and T are said to be *skew* if $S \wedge T = \varnothing$.

LEMMA 11.14. The set-theoretic intersection of any number of subspaces is a subspace.

The proof is left as an exercise.

Let S and T be subspaces. Then S and T are on some subspace; in fact, if $S = \langle A_1, \ldots, A_m \rangle$ and $T = \langle B_1, \ldots, B_n \rangle$, then $S, T \circ \langle A_1, \ldots, A_m, B_1, \ldots, B_n \rangle$. Now the set-theoretic intersection of all subspaces containing S and T is a subspace and will be called the *join of* S *and* T. We shall write $S \vee T$ for the join of S and T. This definition of join is in agreement with our earlier definition when one of S or T is a point. Note, also, that $S \vee T$ is the smallest (in the set-inclusion sense) subspace containing both S and T. Finally, for $A_1, \ldots, A_m, B_1, \ldots, B_n$ points,

$$\langle A_1, \ldots, A_m \rangle \vee \langle B_1, \ldots, B_n \rangle = \langle A_1, \ldots, A_m, B_1, \ldots, B_n \rangle.$$

We now give an alternate characterization of $S \vee T$.

THEOREM 11.15. Let S, T be subspaces. Then C is a point on $S \vee T$ if and only if there exist a point A on S and a point B on T such that $C \circ A \vee B$.

PROOF. For the "only if" part, the proof will be inductive. Let $S = \langle A_1, \ldots, A_m \rangle$. Now let $T = \langle B_1 \rangle$ and $C \circ S \vee T$. Then $S \vee T = \langle A_1, \ldots, A_m, B_1 \rangle$ and, by Lemma 11.3, there exists $A \circ S$ such that

$C \circ B_1 \vee A = A \vee B_1$. Hence the statement holds for $n = 1$. Let the statement hold for $n = k$, and let $T = \langle B_1, \ldots, B_{k+1} \rangle$ and $C \circ S \vee T$. Then $S \vee T = \langle A_1, \ldots, A_m, B_1, \ldots, B_{k+1} \rangle$ and there exists $D \circ \langle A_1, \ldots, A_m, B_1, \ldots, B_k \rangle$ such that $C \circ B_{k+1} \vee D$ (see Figure 11.4).

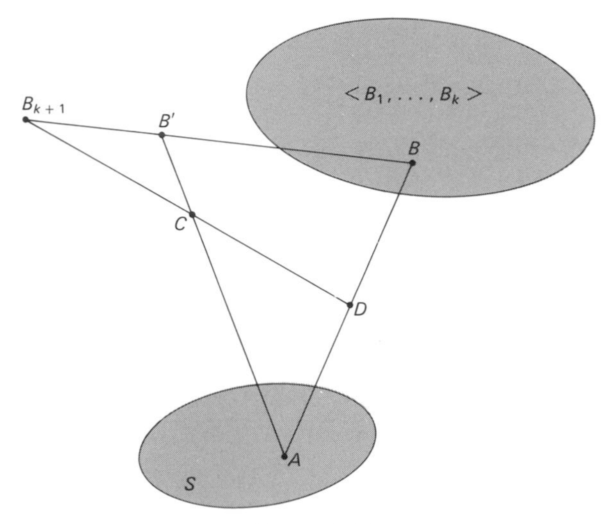

Figure 11.4

By the induction hypothesis, there exist $A \circ S$ and $B \circ \langle B_1, \ldots, B_k \rangle$ such that $D \circ A \vee B$. Consider the case when A, B, B_{k+1} are distinct and noncollinear. (The remaining cases are left as an exercise.) Then lines $B_{k+1} \vee B$ and $A \vee C$, being on $A \vee B \vee B_{k+1}$, intersect in a point B'. Now $C \circ A \vee B'$ and $B' \circ \langle B_1, \ldots, B_{k+1} \rangle$. Hence the statement holds for $n = k + 1$. This completes the "only if" part. The converse is left for the exercise.

Our next result is presented here to be used in the proof of the result following it. However, the result is important in its own right when the theory of projective spaces is studied from an alternate point of view (using systems called *lattices*) and is usually called the *Modular Law*.

THEOREM 11.16. If R, S, T are subspaces such that $R \circ T$, then $(R \vee S) \wedge T = R \vee (S \wedge T)$.

This property expresses a certain "nonparallelism" of our geometry, which can be illustrated in the following ways:

(a) Let S and T be lines such that $S \vee T$ is a plane and R a point not on S.
(b) Let S and T be planes such that $S \vee T$ has dimension 3 and $R \not\circ S$, $R \neq T$.
(c) Let S be a line and T a plane such that $S \vee T$ has dimension 3 and $R \not\circ S$, $R \neq T$.

The details of the illustrations are left as an exercise.

PROOF OF THEOREM 11.16. Let R, S, T be subspaces with $R \circ T$. For the inclusion $(R \vee S) \wedge T \subset R \vee (S \wedge T)$, consider $R \not\circ S$. (The case $R \circ S$ is left as an exercise.) Let A be a point on $(R \vee S) \wedge T$. Then $A \circ R \vee S$ and $A \circ T$. Further, by Theorem 11.15, there exist a point $B \circ R$ and a point $C \circ S$ such that $A \circ B \vee C$. In case $A = B$, we have $A \circ R$. Hence $A \circ R \vee (S \wedge T)$. Let $A \neq B$. Then $C \circ A \vee B$. Therefore $C \circ T$ since $A \circ T$ and $B \circ T$. Hence $C \circ S \wedge T$. From this, we have $A \circ R \vee (S \wedge T)$. Thus $(R \vee S) \wedge T \subset R \vee (S \wedge T)$. The reverse inclusion is left an an exercise.

THEOREM 11.17. If $\{A_1, \ldots, A_j, B_1, \ldots, B_m\}$ and $\{A_1, \ldots, A_j, C_1, \ldots, C_n\}$ are independent sets of points such that $\langle A_1, \ldots, A_j, B_1, \ldots, B_m\rangle \wedge \langle A_1, \ldots, A_j, C_1, \ldots, C_n\rangle = \langle A_1, \ldots, A_j\rangle$, then $\{A_1, \ldots, A_j, B_1, \ldots, B_m, C_1, \ldots, C_n\}$ is independent.

We intend that an interpretation for $j = 0$ be included in the theorem, namely, $\langle A_1, \ldots, A_j\rangle = \varnothing$. The proof will be valid for this interpretation.

PROOF OF THEOREM 11.17. Let $\{A_1, \ldots, A_j, B_1, \ldots, B_m\}$ be independent and $S = \langle A_1, \ldots, A_j, B_1, \ldots, B_m\rangle$. We shall proceed with induction on n. Let $\{A_1, \ldots, A_j, C_1\}$ be independent such that $S \wedge \langle A_1, \ldots, A_j, C_1\rangle = \langle A_1, \ldots, A_j\rangle$. Then $C_1 \not\circ S$; otherwise, if $C_1 \circ S$, then $C_1 \circ \langle A_1, \ldots, A_j\rangle$ contrary to hypothesis. Now suppose $A_1 \circ$

$\langle A_2, \ldots, A_j, B_1, \ldots, B_m, C_1\rangle$. Then by Theorem 11.5, $\langle A_2, \ldots, A_j, B_1, \ldots, B_m, C_1\rangle = S$ since $A_1 \not\circ \langle A_2, \ldots, A_j, B_1, \ldots, B_m\rangle$. Hence $C_1 \not\circ S$, a contradiction. Therefore $A_1 \not\circ \langle A_2, \ldots, A_j, B_1, \ldots, B_m, C_1\rangle$. Similarly, $A_i \not\circ \langle A_1, \ldots, A_{i-1}, A_{i+1}, \ldots, A_j, B_1, \ldots, B_m, C_1\rangle$ for $1 \leqq i \leqq j$, and $B_i \not\circ \langle A_1, \ldots, A_j, B_1, \ldots, B_{i-1} B_{i+1}, \ldots, B_m, C_1\rangle$ for $1 \leqq i \leqq m$. Thus $\{A_1, \ldots, A_j, B_1, \ldots, B_m, C_1\}$ is independent and the statement holds for $n = 1$.

Let the statement hold for $n = k$; let $\{A_1, \ldots, A_j, C_1, \ldots, C_{k+1}\}$ be independent such that $S \wedge \langle A_1, \ldots, A_j, C_1, \ldots, C_{k+1}\rangle = \langle A_1, \ldots, A_j\rangle$. Then $\langle A_1, \ldots, A_j, B_1, \ldots, B_m\rangle \wedge \langle A_1, \ldots, A_j, C_1, \ldots, C_k\rangle = \langle A_1, \ldots, A_j\rangle$ since

$$\begin{aligned}\langle A_1, \ldots, A_j\rangle &\subset S \wedge \langle A_1, \ldots, A_j, C_1, \ldots, C_k\rangle\\ &\subset S \wedge \langle A_1, \ldots, A_j, C_1, \ldots, C_{k+1}\rangle\\ &= \langle A_1, \ldots, A_j\rangle.\end{aligned}$$

Now by the induction hypothesis, $\{A_1, \ldots, A_j, B_1, \ldots, B_m, C_1, \ldots, C_k\}$ is independent since $\{A_1, \ldots, A_j, C_1, \ldots, C_k\}$ is independent. Further, using the Modular Law, we obtain

$$\begin{aligned}&\langle A_1, \ldots, A_j, C_1, \ldots, C_k\rangle\\ &\quad= \langle C_1, \ldots, C_k\rangle \vee \langle A_1, \ldots, A_j\rangle\\ &\quad= \langle C_1, \ldots, C_k\rangle \vee (S \wedge \langle A_1, \ldots, A_j, C_1, \ldots, C_{k+1}\rangle)\\ &\quad= (\langle C_1, \ldots, C_k\rangle \vee S) \wedge \langle A_1, \ldots, A_j, C_1, \ldots, C_{k+1}\rangle\\ &\quad= \langle A_1, \ldots, A_j, B_1, \ldots, B_m, C_1, \ldots, C_k\rangle \wedge\\ &\qquad\quad \langle A_1, \ldots, A_j, C_1, \ldots, C_{k+1}\rangle.\end{aligned}$$

Hence $C_{k+1} \not\circ \langle A_1, \ldots, A_j, B_1, \ldots, B_m, C_1, \ldots, C_k\rangle$; otherwise, if $C_{k+1} \circ \langle A_1, \ldots, A_j, B_1, \ldots, B_m, C_1, \ldots, C_k\rangle$, we should have $C_{k+1} \circ \langle A_1, \ldots, A_j, C_1, \ldots, C_k\rangle$ contrary to hypothesis. Now suppose $A_1 \circ \langle A_2, \ldots, A_j, B_1, \ldots, B_m, C_1, \ldots, C_{k+1}\rangle$. Then by Theorem 11.5,

$$\langle A_2, \ldots, A_j, B_1, \ldots, B_m, C_1, \ldots, C_{k+1}\rangle = \langle A_1, \ldots, A_j, B_1, \ldots, B_m, C_1, \ldots, C_k\rangle$$

since $A_1 \not\circ \langle A_2, \ldots, A_j, B_1, \ldots, B_m, C_1, \ldots, C_k\rangle$. Hence $C_{k+1} \circ \langle A_1, \ldots, A_j, B_1, \ldots, B_m, C_1, \ldots, C_k\rangle$ contrary to the above. Therefore $A_1 \not\circ \langle A_2, \ldots, A_j, B_1, \ldots, B_m, C_1, \ldots, C_{k+1}\rangle$. Similarly,

$A_i \not\in \langle A_1, \ldots, A_{i-1}, A_{i+1}, \ldots, A_j, B_1, \ldots, B_m, C_1, \ldots, C_{k+1} \rangle$ for $1 \leqq i \leqq j$;

$B_i \not\in \langle A_1, \ldots, A_j, B_1, \ldots, B_{i-1}, B_{i+1}, \ldots, B_m, C_1, \ldots, C_{k+1} \rangle$ for $1 \leqq i \leqq m$;

$C_i \not\in \langle A_1, \ldots, A_j, B_1, \ldots, B_m, C_1, \ldots, C_{i-1}, C_{i+1}, \ldots, C_{k+1} \rangle$ for $1 \leqq i \leqq k$.

Hence $\{A_1, \ldots, A_j, B_1, \ldots, B_m, C_1, \ldots, C_{k+1}\}$ is independent. Thus the statement holds for $n = k + 1$. This completes the proof.

THEOREM 11.18. If S and T are subspaces, then $\dim S + \dim T = \dim(S \vee T) + \dim(S \wedge T)$.

PROOF. Let S, T be subspaces and let $\{A_1, \ldots, A_j\}$ be a basis of $S \wedge T$, where the A_i are distinct and $j = 0$ if $S \wedge T = \varnothing$. Then there exist a basis $\{A_1, \ldots, A_j, B_1, \ldots, B_m\}$ of S with the B_i distinct and different from the A_i and a basis $\{A_1, \ldots, A_j, C_1, \ldots, C_n\}$ of T with the C_i distinct and different from the A_i. Consider $m = 0$ if $S \circ T$ and $n = 0$ if $T \circ S$. Now $R = \{A_1, \ldots, A_j, B_1, \ldots, B_m, C_1, \ldots, C_n\}$ spans $S \vee T$ and, by Theorem 11.17, R is independent. Hence R is a basis of $S \vee T$. Now $\dim S = j + m - 1$, $\dim T = j + n - 1$, $\dim(S \vee T) = j + m + n - 1$, and $\dim(S \wedge T) = j - 1$, and the desired equality follows readily.

EXERCISES

11.1. Complete the proof of Lemma 11.1.

11.2. Show that

(a) $\{A_1, \ldots, A_n\}$ spans $\langle A_1, \ldots, A_n \rangle$ for $A_1, \ldots, A_n$ points;

(b) $\varnothing$ spans $\varnothing$;

(c) $\{\{P\}: P \in \mathscr{P}\}$ spans $\mathscr{P}$;

(d) If S spans T, then $A \in S$ implies $A \subset T$.

11.3. Show that if $\{A_1, \ldots, A_m\} \subset \{B_1, \ldots, B_n\}$ for $A_1, \ldots, A_m, B_1, \ldots, B_n$ points, then $\langle A_1, \ldots, A_m \rangle \circ \langle B_1, \ldots, B_n \rangle$.

11.4. Prove Lemma 11.3.

11.5. Complete the proofs of Lemma 11.4 and Theorems 11.5 and 11.6.

11.6. Prove Theorem 11.7.

11.7. Show that

(a) $\varnothing$ is independent;

(b) for A, B points, $\{A\}$ and $\{A, B\}$ are independent;

(c) any subset of an independent set is independent;

(d) $A_i \circ \langle A_1, \ldots, A_{i-1}, A_{i+1}, \ldots, A_n \rangle$ for some i is necessary but not sufficient for $\{A_1, \ldots, A_n\}$ to be a dependent set of points.

11.8. Complete the proof of Theorem 11.10.

11.9. Prove Corollary 11.12.

11.10. Prove Lemma 11.14.

11.11. Show that for S, T subspaces, $S \vee T$ is the smallest subspace containing S and T. Show also that for A a point, $A \vee S$ is the smallest subspace containing both A and S. Hence the new meaning of join agrees with the old.

11.12. Show that for $A_1, \ldots, A_m, B_1, \ldots, B_n$ points, $\langle A_1, \ldots, A_m \rangle \vee \langle B_1, \ldots, B_n \rangle = \langle A_1, \ldots, A_m, B_1, \ldots, B_n \rangle$.

11.13. Let $(\mathscr{P}, \mathscr{L}, \circ)$ be n-dimensional. Show that

(a) if $n = 3$, every line and plane intersect;

(b) if $n \geqq 3$, there exist skew lines;

(c) if $n \geqq 4$, there exist a skew line and plane;

(d) if $n = 4$, every line intersects every subspace having dimension 3;

(e) if $n = 4$, there are no skew planes;

(f) if $n \geqq 5$, there exist skew planes.

11.14. Show that

(a) Basis 10.1 is $(n - 1)$-dimensional;

(b) Basis 10.2 does not have finite dimension if the set S is not finite.

11.15. Complete the proof of Theorem 11.15.

11.16. Supply the details of the suggested "nonparallelism" illustrations following the statement of Theorem 11.16.

11.17. Complete the proof of Theorem 11.16.

11.18. Prove the converse of Theorem 11.17.

11.2. INTERVALS AND COMPLEMENTS

Throughout this section we shall assume that $(\mathscr{P}, \mathscr{L}, \circ)$ is an n-dimensional projective space where $n \geqq 2$. Our results will shed some light on the structure of the elements of $\overline{\mathscr{P}}$ relative to the incidence relation $\circ$ and our operations $\vee$ and $\wedge$.

Let S and T be subspaces such that $S \circ T$. The *interval from S to T*, written $[S, T]$, is defined to be

$$\{X \in \overline{\mathscr{P}} : S \circ X \quad \text{and} \quad X \circ T\}.$$

The *length* of $[S, T]$ is defined to be $\dim T - \dim S + 1$. We have immediately that the length of $[\varnothing, T] = \dim T + 2$.

THEOREM 11.19. If R, S, T are subspaces such that $R \circ S$ and $S \circ T$, then there exists a subspace S' such that $S \wedge S' = R$ and $S \vee S' = T$.

The subspace S' of the theorem will be called a (*relative*) *complement of S in* $[R, T]$. In case $R = \varnothing$ and $T = \mathscr{P}$, we shall simply call S' a *complement* of S.

PROOF OF THEOREM 11.19. Let $R \circ S$, $S \circ T$ be subspaces and let $\{A_1, \ldots, A_j\}$ be a basis of R with the A_i distinct. Then there exist a basis $\{A_1, \ldots, A_j, B_1, \ldots, B_k\}$ of S such that $\dim S = j + k - 1$ and a basis $\{A_1, \ldots, A_j, B_1, \ldots, B_k, C_1, \ldots, C_m\}$ of T such that $\dim T = j + k + m - 1$. Define $S' = \langle A_1, \ldots, A_j, C_1, \ldots, C_m \rangle$. Then $S \vee S' = T$. Now $R \circ S \wedge S'$ and

$$\begin{aligned}
\dim(S \wedge S') &= \dim S + \dim S' - \dim(S \vee S') \\
&= (j + k - 1) + (j + m - 1) - (j + k + m - 1) \\
&= j - 1 \\
&= \dim R.
\end{aligned}$$

Thus $S \wedge S' = R$.

In part (c) of Corollary 11.12, the elements of a subspace S having dimension $k \geqq 2$ are shown to form a k-dimensional projective space where $(\mathbf{P}, \mathbf{L}, \odot)$ is the incidence basis considered if $\mathbf{P} = \{A \in \overline{\mathscr{P}} : A \circ S$ and $\dim A = 0\}$, $\mathbf{L} = \{L \in \overline{\mathscr{P}} : L \circ S$ and $\dim L = 1\}$, and for $B \in \mathbf{P}$ and $M \in \mathbf{L}$, $B \odot M$ means $B \circ M$. Note that $\mathbf{P}$ and $\mathbf{L}$ are equivalently given by

$$\mathbf{P} = \{A \in [\varnothing, S] : \dim A = 0\}$$

and

$$\mathbf{L} = \{L \in [\varnothing, S] : \dim L = 1\}.$$

The idea expressed in part (c) of corollary 11.12 can be extended to any interval in the following way. Let S, T be distinct subspaces such that $S \circ T$. Define

$$\mathbf{P} = \{A \in [S, T] : \dim A = \dim S + 1\},$$
$$\mathbf{L} = \{L \in [S, T] : \dim L = \dim S + 2\},$$

and for $B \in \mathbf{P}$ and $M \in \mathbf{L}$, define $B \odot M$ to mean $B \circ M$. We shall call $(\mathbf{P}, \mathbf{L}, \odot)$ the incidence basis *associated* with the interval $[S, T]$.

LEMMA 11.20. If S, T are distinct subspaces such that $S \circ T$ and S' is a complement of S in $[\varnothing, T]$, then the incidence bases associated with $[S, T]$ and $[\varnothing, S']$ are isomorphic.

PROOF. Let S, T be distinct subspaces such that $S \circ T$ and let S' be a complement of S in $[\varnothing, T]$. Further, let $(\mathbf{P}, \mathbf{L}, \odot)$ and $(\mathbf{P}', \mathbf{L}', \odot')$ be the incidence bases associated with $[S, T]$ and $[\varnothing, S']$, respectively. The result is immediate if $\dim T - \dim S = 1$. Let $\dim T - \dim S \geqq 2$. Define

$$\pi A = A \wedge S' \qquad \text{for} \quad A \in \mathbf{P}$$

and

$$\lambda L = L \wedge S' \qquad \text{for} \quad L \in \mathbf{L}.$$

For $A' \in \mathbf{P}'$, we have $A' \vee S \in \mathbf{P}$ and, by the Modular Law, $(A' \vee S) \wedge S' = A' \in P'$. It is left as an exercise to show that π is a one-to-one correspondence between $\mathbf{P}$ and $\mathbf{P}'$ and that $\pi^{-1}A' = A' \vee S$ for $A' \in \mathbf{P}'$. Left also as part of the exercise is the proof that λ is a one-to-one correspondence between $\mathbf{L}$ and $\mathbf{L}'$ and that $\lambda^{-1}L' = L' \vee S$ for $L' \in \mathbf{L}'$. Now let $A \in \mathbf{P}$ and $L \in \mathbf{L}$. If $A \odot L$, then $A \wedge S' \odot L \wedge S'$, so that $\pi A \odot' \lambda L$. Conversely, for $\pi A \odot' \lambda L$, we have $A = \pi^{-1}\pi A = \pi A \vee S \odot \lambda L \vee S = \lambda^{-1}\lambda L = L$. Thus (π, λ) is an isomorphism between $(\mathbf{P}, \mathbf{L}, \odot)$ and $(\mathbf{P}', \mathbf{L}', \odot')$.

LEMMA 11.21. If S_1 and S_2 are subspaces with the same dimension, then they have a common complement in $[S_1 \wedge S_2, S_1 \vee S_2]$.

PROOF. Let $S_1, S_2 \in \overline{\mathscr{P}}$ such that $S_1 \neq S_2$ and $\dim S_1 = \dim S_2$. Define $S = S_1 \wedge S_2$ and $T = S_1 \vee S_2$. In light of Lemma 11.20, it is sufficient to consider $S = \varnothing$. (Let S' be a complement of S in $[\varnothing, T]$ and transfer the study of $[S, T]$ to a study of $[\varnothing, S']$.) The proof will proceed with induction on the dimension of the subspaces S_1 and S_2. For $\dim S_1 = 0$, S_1 and S_2 are points and a third point on line T will serve as the common complement in $[\varnothing, T]$. Thus the statement holds for $\dim S_1 = 0$. Let the statement hold when the dimension of the subspaces is $k \geqq 0$ and let $\dim S_1 = k + 1$. Now let $T_1 \in [S_1, T]$ and $T_2 \in [S_2, T]$ such that $\dim T_1 = \dim T_2 = \dim T - 1$. (How can we establish the existence of such subspaces?) Define $T' = T_1 \wedge T_2$ and $S_1' = S_1 \wedge T'$, $S_2' = S_2 \wedge T'$. Then $\dim T' = \dim T - 2$ and $\dim S_1' = \dim S_2' = k$. (Why?) In addition, $S_1' \wedge S_2' = \varnothing$ and $S_1' \vee S_2' = T'$. By the induction hypothesis, there exists $C' \in \overline{\mathscr{P}}$ such that C' is a complement of S_1' and S_2' in $[\varnothing, T']$ (see Figure 11.5 which illustrates the

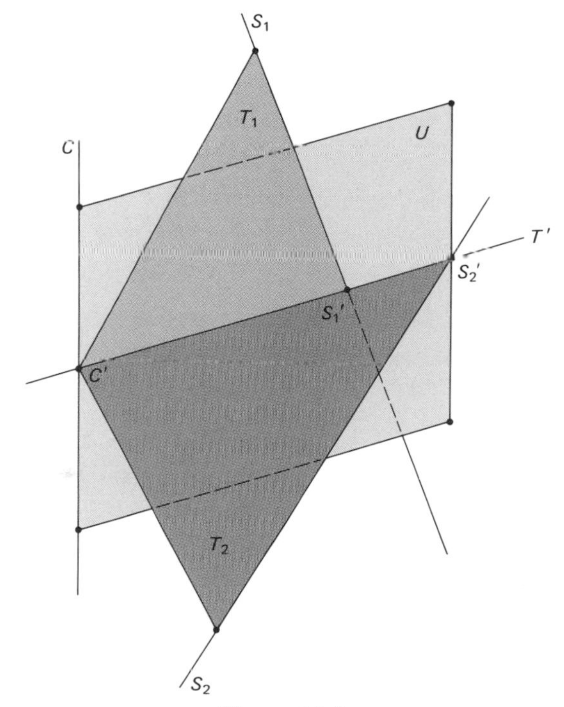

Figure 11.5

proof for the case $\dim S_1 = 1$ and $\dim T = 3$). Then $\dim C' = k$. Now we apply the case already established ($\dim S_1 = 0$) and Lemma 11.20 to obtain a complement U of T_1 and T_2 in $[T', T]$. Then $\dim U = \dim T - 1$. Finally, let C be a complement of T' in $[C', U]$. Then $\dim C = k + 1$. To show $C \vee S_1 = T$, let $D = C \vee S_1$. Then $D \circ T$. In addition, $T' = S_1' \vee C'$ is on D since S_1', $C \circ D$, but $T' \neq D$ since $C \circ D$ and $C \not\circ T'$. Hence $\dim D = \dim T$ or $\dim D = \dim T - 1$. Suppose $\dim D = \dim T - 1$. (This means C and S_1 are coplanar in Figure 11.5.) Then $T_1 = S_1 \vee T'$ is on D since S_1, $T' \circ D$. Therefore by Corollary 11.12, $T_1 = D$. Similarly, $U = C \vee T' \circ D$, so that $U = D$. Hence $T_1 = U$ contrary to the hypothesis on U. Thus $\dim D = \dim T$, so that $D = T$. Now by Theorem 11.18, $C \wedge S_1 = \varnothing$. Thus C is a complement of S_1 in $[\varnothing, T]$. We can show, similarly, that C is a complement of S_2 in $[\varnothing, T]$.

THEOREM 11.22. If S_1 and S_2 are nonempty subspaces with the same dimension, then the incidence bases associated with $[\varnothing, S_1]$ and $[\varnothing, S_2]$ are isomorphic.

PROOF. Let S_1 and S_2 be nonempty subspaces with the same dimension. Now let S' be a complement of S_1 and S_2 in $[S_1 \wedge S_2, S_1 \vee S_2]$. Finally, let S be a complement of $S_1 \wedge S_2$ in $[\varnothing, S']$. Then S_1 and S_2 are complements of S in $[\varnothing, S_1 \vee S_2]$. (Why?) Now by Lemma 11.20, the incidence basis associated with $[S, S_1 \vee S_2]$ is isomorphic to the incidence bases associated with $[\varnothing, S_1]$ and $[\varnothing, S_2]$, so that the latter two are isomorphic.

COROLLARY 11.23.

(a) The incidence bases associated with intervals of the same length are isomorphic.
(b) The incidence basis associated with an interval of length k for $k \geqq 4$ is a $(k - 2)$-dimensional projective space.

The proof is left as an exercise.

The complement of a point will be called a *hyperplane* and the complement of a line a *subhyperplane*.

COROLLARY 11.24. Hyperplanes and subhyperplanes have dimension $n - 1$ and $n - 2$, respectively.

The proof is left as an exercise.

EXERCISES

11.19. Show that if S' is a complement of a subspace S, then $\dim S' = n - 1 - \dim S$.
11.20. Complete the proof of Lemma 11.20.
11.21. Complete the proof of Lemma 11.21.
11.22. Complete the proof of Theorem 11.22.
11.23. Prove Corollaries 11.23 and 11.24.

11.3. DUAL SPACES

We shall continue our assumption, in this section, that $(\mathscr{P}, \mathscr{L}, \circ)$ is an n-dimensional projective space. A "Principle of Duality" will not be formulated for projective spaces. Instead, we shall show how the hyperplanes and subhyperplanes of $(\mathscr{P}, \mathscr{L}, \circ)$ can be used, respectively, as "points" and "lines" in the construction of a "dual" n-dimensional projective space. Of course, in case $n = 2$, we have both our original Principle of Duality and this new dual projective plane (see Exercise 3.6).

Define $\mathscr{H} = \{R \in \overline{\mathscr{P}} : R \text{ is a hyperplane}\}$ and $\mathscr{S} = \{R \in \overline{\mathscr{P}} : R \text{ is a subhyperplane}\}$. For $H \in \mathscr{H}$ and $S \in \mathscr{S}$, define $H \mathbin{\bar{\circ}} S$ to mean $S \circ H$. Then $(\mathscr{H}, \mathscr{S}, \bar{\circ})$ is an incidence basis.

THEOREM 11.25. The incidence basis $(\mathscr{H}, \mathscr{S}, \bar{\circ})$ is a projective space.

We define $(\mathscr{H}, \mathscr{S}, \bar{\circ})$ to be the *dual space* of $(\mathscr{P}, \mathscr{L}, \circ)$.

PROOF OF THEOREM 11.25. For Axiom P_1, let L be a line and S be a complement of L. Then $S \neq \varnothing$. Let A be a point on S and H be a complement of A. It is left as an exercise to show that $H \not\mathbin{\bar{\circ}} S$.

For Axiom P_2, let $S \in \mathscr{S}$. Now let L be a complement of S and A, B, C be distinct points on L. That $A \vee S, B \vee S, C \vee S \in \mathscr{H}$ and are distinct is left for the exercise. Thus Axiom P_2 holds.

Now let $H, J \in \mathscr{H}$ such that $H \neq J$. Then $H, J \mathbin{\bar{\circ}} H \wedge J$. By our dimension theory, $H \wedge J \in \mathscr{S}$, and $H \wedge J$, with these properties, is unique. Thus Axiom P_3 holds and we have $H \wedge J$ as the "join" of H and J in $(\mathscr{H}, \mathscr{S}, \bar{\circ})$.

Finally, for Axiom P_4*, let $H, I, J, K \in \mathscr{H}$ and be distinct and noncollinear such that the "join" of H, I and the "join" of J, K "intersect", that is, $H \wedge I, K \wedge J \circ G$ for some $G \in \mathscr{H}$. Let $S = H \wedge I$ and $T = K \wedge J$. Then $S \vee T = G$, so that $\dim(S \wedge T) = n - 3$. Let $R = S \wedge T$ and R' be a complement of R. Then R' is a plane. (Why?) In addition, $H' = H \wedge R'$, $I' = I \wedge R'$, $J' = J \wedge R'$, and $K' = K \wedge R'$ are distinct nonconcurrent lines on R'. (Why?) Define $F' = (H' \wedge K') \vee (I' \wedge J')$ and $F = F' \vee R$. Then F' is a line on R'. Hence $F \in \mathscr{H}$. Further, $H \wedge K, I \wedge J \circ F$, so that F is the "intersection" of $H \wedge K$ and $I \wedge J$ in $(\mathscr{H}, \mathscr{S}, \bar{\circ})$. This completes the proof of Theorem 11.25.

We are, now, in a position to use our generalized join and independence theory for the "points" of $(\mathscr{H}, \mathscr{S}, \bar{\circ})$.

LEMMA 11.26. For $H, J, K \in \mathscr{H}$, $H \in \langle\{J\}, \{K\}\rangle$ if and only if $J \wedge K \circ H$.

The proof is left as an exercise.

LEMMA 11.27. For $H_1, \ldots, H_m, H \in \mathscr{H}$, $H \in \langle\{H_1\}, \ldots, \{H_m\}\rangle$ if and only if $H_1 \wedge \cdots \wedge H_m \circ H$.

PROOF. We shall proceed inductively. The statement obviously holds for $m = 1$. Let the statement hold for $m = k$, and let $H_1, \ldots, H_{k+1}, H \in \mathscr{H}$. For the "only if" part, let $H \in \langle\{H_1\}, \ldots, \{H_{k+1}\}\rangle$. Then H is "on" the "join" of H_{k+1} and J for some $J \in \langle\{H_1\}, \ldots, \{H_k\}\rangle$, that is, $H_{k+1} \wedge J \circ H$. Now, by the induction hypothesis, $H_1 \wedge \cdots \wedge H_k \circ J$. Thus $H_1 \wedge \cdots \wedge H_{k+1} \circ H$. For the converse, let $H_1 \wedge \cdots \wedge H_{k+1} \circ H$. Let $K = H_1 \wedge \cdots \wedge H_k$. Consider the case when $K \not\circ H$; otherwise, we have by the induction hypothesis,

$$H \in \langle\{H_1\}, \ldots, \{H_k\}\rangle \subset \langle\{H_1\}, \ldots, \{H_{k+1}\}\rangle.$$

Now $K \not\circ H_{k+1}$. Let $S = H_{k+1} \wedge H$ and $H' = H \vee S$. Then $H' \in \mathscr{H}$ and $K \circ H'$. Hence by the induction hypothesis, $H' \in \langle\{H_1\}, \ldots, \{H_k\}\rangle$. Further, $H' \wedge H_{k+1} = S \circ H$, so that by Lemma 11.26, $H \subset \langle\{H'\}, \{H_{k+1}\}\rangle$. Hence $H \in \langle\{H_1\}, \ldots, \{H_{k+1}\}\rangle$. Thus the statement holds for $m = k + 1$. This completes the proof.

THEOREM 11.28. Let S be a subspace in $(\mathscr{P}, \mathscr{L}, \circ)$. Then $\dim S = m$ if and only if there exist distinct $H_1, \ldots, H_{n-m} \in \mathscr{H}$ such that $S = H_1 \wedge \cdots \wedge H_{n-m}$ and $\{\{H_1\}, \ldots, \{H_{n-m}\}\}$ is independent in $(\mathscr{H}, \mathscr{S}, \bar{\circ})$.

PROOF. Let S be a subspace. For the "only if" part, let $\{A_1, \ldots, A_{m+1}\}$ be a basis of S and $\{A_1, \ldots, A_{m+1}, B_1, \ldots, B_{n-m}\}$ be a basis of $\mathscr{P}$. Define

$$H_i = \langle A_1, \ldots, A_{m+1}, B_1, \ldots, B_{i-1}, B_{i+1}, \ldots, B_{n-m}\rangle$$

for $i = 1, \ldots, n - m$. Then $H_i \in \mathscr{H}$ and $S = H_1 \wedge \cdots \wedge H_{n-m}$. Further, $H_i \neq H_j$ for $i \neq j$. Suppose $H_1 \in \langle\{H_2\}, \ldots, \{H_{n-m}\}\rangle$. Then $H_2 \wedge \cdots \wedge H_{n-m} \circ H_1$. But $B_1 \circ H_2 \wedge \cdots \wedge H_{n-m}$, so that $B_1 \circ H_1$ contrary to the definition of H_1. Hence $H_1 \notin \langle\{H_2\}, \ldots, \{H_{n-m}\}\rangle$. Similarly $H_i \notin \langle\{H_1\}, \ldots, \{H_{i-1}\}, \{H_{i+1}\}, \ldots, \{H_{n-m}\}\rangle$ for $i = 1, \ldots, n - m$. Thus $\{\{H_1\}, \ldots, \{H_{n-m}\}\}$ is independent. The converse is left as an exercise.

COROLLARY 11.29. If S is a subspace of $(\mathscr{P}, \mathscr{L}, \circ)$ having dimension m, then $\{H \in \mathscr{H} : S \circ H\}$ is a subspace of $(\mathscr{H}, \mathscr{S}, \bar{\circ})$ having dimension $n - m - 1$. In particular, $(\mathscr{H}, \mathscr{S}, \bar{\circ})$ is n-dimensional.

The proof is left as an exercise.

We shall conclude this chapter with the result that the dual space of the dual space of $(\mathscr{P}, \mathscr{L}, \circ)$ brings us back to $(\mathscr{P}, \mathscr{L}, \circ)$ (isomorphically, that is). An element $P \in \mathscr{P}$ will correspond to $\langle\{H_1\}, \ldots, \{H_n\}\rangle$, a hyperplane in $(\mathscr{H}, \mathscr{S}, \bar{\circ})$, where $H_1, \ldots, H_n \in \mathscr{H}$ such that $\{P\} = H_1 \wedge \cdots \wedge H_n$. We shall find the next lemma useful in the proof of our concluding theorem.

LEMMA 11.30. Let $P, Q, R \in \mathscr{P}$, $P_i, Q_i, R_i \in \mathscr{H}$ such that $\{P\} = P_1 \wedge \cdots \wedge P_n, \{Q\} = Q_1 \wedge \cdots \wedge Q_n$, $\{R\} = R_1 \wedge \cdots \wedge R_n$ and $\bar{P} = \langle\{P_1\}, \ldots, \{P_n\}\rangle$, $\bar{Q} = \langle\{Q_1\}, \ldots, \{Q_n\}\rangle$, $\bar{R} = \langle\{R_1\}, \ldots, \{R_n\}\rangle$. Then P, Q, R are collinear in $(\mathscr{P}, \mathscr{L}, \circ)$ if and only if $\bar{P}, \bar{Q}, \bar{R}$ are collinear in the dual space of $(\mathscr{H}, \mathscr{S}, \bar{\circ})$.

PROOF. Let $P, Q, R, P_i, Q_i, R_i, \bar{P}, \bar{Q}, \bar{R}$ be as stated in the hypothesis of the lemma. Then $\bar{P}, \bar{Q}, \bar{R}$ are collinear "points" in the dual space of $(\mathscr{H}, \mathscr{S}, \bar{\circ})$ if and only if there exists a subhyperplane $\bar{S}$ of $(\mathscr{H}, \mathscr{S}, \bar{\circ})$ such that $\bar{S} \subset \bar{P}, \bar{Q}, \bar{R}$, that is, $\bar{S} = \bar{P} \cap \bar{Q} \cap \bar{R}$ and $\dim \bar{S} = n - 2$. The set $\bar{S}$ consists of the hyperplanes of $(\mathscr{P}, \mathscr{L}, \circ)$ that $\{P\}, \{Q\}, \{R\}$ are on, that is, the hyperplanes that $\langle\{P\}, \{Q\}, \{R\}\rangle$ is on. By Corollary 11.29, $\dim \bar{S} = n - \dim \langle\{P\}, \{Q\}, \{R\}\rangle - 1$. Hence $\dim \bar{S} = n - 2$ if and only if $\dim \langle\{P\}, \{Q\}, \{R\}\rangle = 1$. The latter holds if and only if $\{P\} \circ \{Q\} \vee \{R\}$, that is, $P \circ Q \vee R$. This completes the proof.

THEOREM 11.31. If $(\mathscr{P}_1, \mathscr{L}_1, \circ_1)$ is the dual space of $(\mathscr{H}, \mathscr{S}, \bar{\circ})$, then $(\mathscr{P}, \mathscr{L}, \circ)$ is isomorphic to $(\mathscr{P}_1, \mathscr{L}_1, \circ_1)$.

PROOF. Let $(\mathscr{P}_1, \mathscr{L}_1, \circ_1)$ be the dual space of $(\mathscr{H}, \mathscr{S}, \bar{\circ})$. Now for $P \in \mathscr{P}$ and $H_1, \ldots, H_n \in \mathscr{H}$ such that $\{P\} = H_1 \wedge \cdots \wedge H_n$, define $\pi P = \langle\{H_1\}, \ldots, \{H_n\}\rangle$. Then $\pi P \in \mathscr{P}_1$ for $P \in \mathscr{P}$. It is left as an exercise to show that π is a one-to-one correspondence between $\mathscr{P}$ and $\mathscr{P}_1$. That π leads to the desired isomorphism follows from the lemma.

EXERCISES

11.24. Complete the proof of Theorem 11.25.
11.25. Prove Lemma 11.26.
11.26. Complete the proof of Theorem 11.28.
11.27. Prove Corollary 11.29.
11.28. Complete the proof of Theorem 11.31.

HILBERT'S AXIOMS FOR A EUCLIDEAN PLANE[†]

APPENDIX A

Hilbert employs five undefined concepts, *point*, *line*, *incidence*, *betweenness* and *congruence*. In stating his axioms, he arranges them into five groups, each group expressing, by itself, certain related intuitive ideas.

GROUP I. AXIOMS OF CONNECTION

In this group, *point*, *line*, and *incidence* are introduced. The points and lines are the objects of study and incidence is a relation between points and lines. We say point P "is incident with" line l. We may also say P "lies on" l, P "is on" l, P "is a point of" l, l "goes through" P, l "passes through" P, and so on.

I–1. Through any pair of distinct points A and B, there is at least one line l.

[†] This is an adaptation from Hilbert's "The Foundations of Geometry" by permission of the Open Court Publishing Co., La Salle, Illinois.

I–2. Through any pair of distinct points A and B, there is at most one line l.

I–3. There exist at least two points on every line.

I–4. There exist at least three points not on the same line.

The unique line l of Axioms I-1 and I-2 will be called *line AB* for A and B distinct points.

GROUP II. AXIOMS OF ORDER

The concept of *betweenness* is introduced in this group. This describes the relationship that three points of a line bear to one another and we express this by saying point B "is between" points A and C.

II–1. If point B is between points A and C, then A, B, C are distinct points on some line, and B is between C and A.

II–2. For any distinct points A and C, there is at least one point B on line AC such that C is between A and B.

II–3. If A, B, C are distinct points on some line, then at most one of the points is between the other two.

For distinct points A and B, the *segment AB* is defined to be the set of all points which are between A and B. The points A and B are called the *endpoints* of segment AB. The segment BA is the same as segment AB.

II–4. (Pasch's Axiom). Let A, B, C be distinct points not on the same line and let l be a line which does not pass through any of the points A, B, C. Then if l passes through a point of the segment AB, it will also pass through a point of segment AC or a point of segment BC.

Let l be a line. Then points A and B, not on l, are said to be on the *same side of l* if l passes through no point of segment AB and on the *opposite side of l* otherwise. The line l, then, divides all points into three mutually exclusive sets, namely, those points on l and those points on one of the two "sides" of l.

For distinct points A and B, the *ray* AB is defined to be the set of points which are between A and B, the point B itself, and all points C such that B is between A and C. The ray AB is said to *emanate from* point A.

It is a consequence of the above axioms that a point A, on a line l, divides l into two rays emanating from A such that two points are on the same ray if and ony if A is not between them.

Let A be a point and b, c rays emanating from A. The set consisting of A, b, and c is called an *angle*. The point A is called the *vertex* of the angle and b, c the *sides*. Let B be a point on ray b and C a point on ray c. Then the angle will be designated by $\sphericalangle BAC$ or $\sphericalangle CAB$.

Let A, B, C be distinct points not on the same line. Then the set consisting of A, B, C and the segments AB, BC, CA is called *triangle ABC*. The three points are called the *vertices* of the triangle; the three segments are called the *sides*, and $\sphericalangle BAC$, $\sphericalangle ABC$, and $\sphericalangle ACB$ are called the *angles*. We say that $\sphericalangle BAC$ is *included* by the sides AB and AC of the triangle.

GROUP III. AXIOM OF PARALLELS

III–1. (Playfair's Postulate). Through a point A not on a line l, there passes at most one line which does not intersect l.

It is on the basis of this axiom and Axiom II-4 that all points can be considered to be "confined" to a "plane."

GROUP IV. AXIOMS OF CONGRUENCE

The concept of *congruence* is a relation between pairs of segments and a relation between pairs of angles. We shall write segment $AB \cong$ segment CD if segment AB "is congruent to" segment CD and $\sphericalangle ABC \cong \sphericalangle DEF$ if $\sphericalangle ABC$ "is congruent to" $\sphericalangle DEF$.

IV–1. If A, B are distinct points, C is a point on a line l and d is one of the two rays on l emanating from C, then there is one and only one point D on d such that segment $CD \cong$ segment AB.

IV–2. If two segments are congruent to a third segment, then they are congruent to each other.

IV–3. If point C is between A and B, and point C' is between A' and B', and if segment $AC \cong$ segment $A'C'$ and segment $CB \cong$ segment $C'B'$, then segment $AB \cong$ segment $A'B'$.

IV–4. If the sides of $\measuredangle BAC$ do not lie on the same line and if A', B' are distinct points, then there is one and only one ray $A'C'$ such that C' is on a given side of line $A'B'$ and $\measuredangle B'A'C' \cong \measuredangle BAC$. Moreover, every angle is congruent to itself.

IV–5. If two sides and the included angle of one triangle are congruent respectively to two sides and the included angle of another triangle, then each of the remaining angles of the first triangle is congruent to the corresponding angle of the second triangle.

GROUP V. AXIOM OF CONTINUITY

V–1. (Dedekind's Axiom of Continuity). For every partition of all the points on a line into two nonempty and nonoverlapping sets such that no point of one set is between two points of the other set, there is a point belonging to one of the sets which is between every other point of that set and every point of the other set.

DIVISION RINGS — APPENDIX B

In the verification that Basis 3.5 is a pappian projective plane, several properties of the real numbers are used, which hold by virtue of the fact that the real number system is a division ring. Hence these properties hold for any division ring. This fact is applied in the proof of Theorem 6.13 without verification. It is the purpose of this appendix to enumerate those properties used that are not explicitly stated among the axioms for a division ring. Some are proved here. The proofs of the remaining ones are left as exercises.

The definition of a division ring is repeated for reference. Let $\mathscr{F}$ be a set and $+$ and $\cdot$ be binary operations on $\mathscr{F} \times \mathscr{F}$ to $\mathscr{F}$. The system $(\mathscr{F}, +, \cdot)$ is called a *division ring* if the following axioms are satisfied:

1. For $a, b, c, \in \mathscr{F}$, $a + (b + c) = (a + b) + c$.
2. There exists an element $0 \in \mathscr{F}$ such that $a + 0 = a$ and $0 + a = a$ for $a \in \mathscr{F}$.
3. For $a \in \mathscr{F}$, there exists an element $b \in \mathscr{F}$ such that $a + b = 0$ and $b + a = 0$.
4. For $a, b \in \mathscr{F}$, $a + b = b + a$.
5. For $a, b, c \in \mathscr{F}$, $a \cdot (b \cdot c) = (a \cdot b) \cdot c$.

6. There exists an element $1 \in \mathscr{F}$ such that $1 \neq 0$, and $a \cdot 1 = a$ and $1 \cdot a = a$ for $a \in \mathscr{F}$.
7. For $a \in \mathscr{F}$ such that $a \neq 0$, there exists an element $b \in \mathscr{F}$ such that $a \cdot b = 1$ and $b \cdot a = 1$.
8. For $a, b, c \in \mathscr{F}$, $a \cdot (b + c) = (a \cdot b) + (a \cdot c)$ and $(a + b) \cdot c = (a \cdot c) + (b \cdot c)$.

We shall call 0 and 1, respectively, *additive* and *multiplicative identities*. For $a \in \mathscr{F}$, we shall call the element b of Axiom 3 an *additive inverse* of a and, in case $a \neq 0$, the element b of Axiom 7 a *multiplicative inverse* of a.

B$_1$. The elements 0 and 1 are unique as additive and multiplicative identities.

PROOF OF B$_1$. Let $x \in \mathscr{F}$ such that $a + x = a$ and $x + a = a$ for $a \in \mathscr{F}$. Then $x = x + 0 = 0$. Thus 0 is unique as the additive identity. The uniqueness of 1 follows similarly.

B$_2$. Additive and multiplicative inverses are unique.

PROOF OF B$_2$. Let $a \in \mathscr{F}$ and let $b, c \in \mathscr{F}$ such that $a + b = 0$, $b + a = 0$ and $a + c = 0$, $c + a = 0$. Then

$$\begin{aligned} b &= b + 0 \\ &= b + (a + c) \\ &= (b + a) + c \\ &= 0 + c \\ &= c. \end{aligned}$$

Thus a has a unique additive inverse. The uniqueness of multiplicative inverses follows similarly.

For $a \in \mathscr{F}$, we shall write $-a$ for the additive inverse of a and, in case $a \neq 0$, a^{-1} for the multiplicative inverse of a.

For $a, b, c \in \mathscr{F}$, we shall write ab for $a \cdot b$, $a + bc$ for $a + (bc)$, and $ab + c$ for $(ab) + c$.

For $a, b, c, d \in \mathscr{F}$, define

$$a + b + c = (a + b) + c,$$
$$a + b + c + d = (a + b + c) + d.$$

Then $a + b + c = a + (b + c)$.

B$_3$. If $a, b, c, d \in \mathscr{F}$, $a(b + c + d) = ab + ac + ad$ and $(a + b + c)d = ad + bd + cd$.

PROOF OF B$_3$. Let $a, b, c, d \in \mathscr{F}$. We shall establish the first equality only. Then

$$\begin{aligned} a(b + c + d) &= a((b + c) + d) \\ &= a(b + c) + ad \\ &= (ab + ac) + ad \\ &= ab + ac + ad. \end{aligned}$$

B$_4$. If $a, b, c, a', b', c' \in \mathscr{F}$,

$$(a + b + c) + (a' + b' + c') = (a + a') + (b + b') + (c + c').$$

PROOF OF B$_4$. Let $a, b, c, a', b', c' \in \mathscr{F}$. Then

$$\begin{aligned} (a + b + c) + (a' + b' + c') &= (a + (b + c)) + (a' + (b' + c')) \\ &= ((b + c) + a) + (a' + (b' + c')) \\ &= (b + c) + (a + (a' + (b' + c'))) \\ &= (b + c) + ((a + a') + (b' + c')) \\ &= ((b + c) + (a + a')) + (b' + c') \\ &= ((a + a') + (b + c)) + (b' + c') \\ &= (a + a') + ((b + c) + (b' + c')) \\ &= (a + a') + ((c + b) + (b' + c')) \\ &= (a + a') + (c + (b + (b' + c'))) \\ &= (a + a') + (c + ((b + b') + c')) \\ &= (a + a') + ((c + (b + b')) + c') \\ &= (a + a') + (((b + b') + c) + c') \\ &= (a + a') + ((b + b') + (c + c')) \\ &= (a + a') + (b + b') + (c + c'). \end{aligned}$$

The next three properties are stated without proof. Property B_4 is useful in their proof.

B$_5$. If $a, b, c, a', b', c', a'', b'', c'' \in \mathscr{F}$,

$$(a + b + c) + (a' + b' + c') + (a'' + b'' + c'') = (a + a' + a'') + (b + b' + b'') + (c + c' + c'').$$

B$_6$. If $a, b, c, d \in \mathscr{F}$, $(a + b) + (c + d) = (a + c) + (b + d)$ and $(a + b) + (c + d) = (a + d) + (c + b)$.

B$_7$. If $a, b, c, a', b', c' \in \mathscr{F}$,

$$(a + b + c) + (a' + b' + c') = (b + a') + (c + c') + (a + b').$$

For $a, b \in \mathscr{F}$, define $a - b = a + (-b)$.

B$_8$. If $a, b \in \mathscr{F}$, $-(a + b) = -a - b$.

PROOF OF B_8. Let $a, b \in \mathscr{F}$. Then

$$\begin{aligned}(a + b) + (-a - b) &= (a + b) + ((-a) + (-b)) \\ &= (a + (-a)) + (b + (-b)) \\ &= 0 + 0 \\ &= 0.\end{aligned}$$

In addition, $(-a - b) + (a + b) = (a + b) + (-a - b) = 0$. Thus by B_2, $-(a + b) = -a - b$.

For $a, b, c, d \in \mathscr{F}$, define

$$\begin{aligned}a + b - c &= a + b + (-c), \\ a - b + c &= a + (-b) + c, \\ a - b - c &= a + (-b) + (-c), \\ a - b + c - d &= a + (-b) + c + (-d).\end{aligned}$$

B$_9$. If $a \in \mathscr{F}$, $a \cdot 0 = 0$ and $0 \cdot a = 0$.

PROOF OF B_9. Let $a \in \mathscr{F}$. We shall establish the first equality only. Then

$$\begin{aligned} a + a \cdot 0 &= a \cdot 1 + a \cdot 0 \\ &= a \cdot (1 + 0) \\ &= a \cdot 1 \\ &= a. \end{aligned}$$

In addition, $a \cdot 0 + a = a + a \cdot 0 = a$. Thus by B_1, $a \cdot 0 = 0$.

B_{10}. If $a, b \in \mathscr{F}$ such that $ab = 0$, then $a = 0$ or $b = 0$.

PROOF OF B_{10}. Let $a, b \in \mathscr{F}$ such that $ab = 0$. Now let $a \neq 0$. Then $b = 1 \cdot b = (a^{-1} \cdot a)b = a^{-1} \cdot (ab) = a^{-1} \cdot 0 = 0$.

B_{11}. If $a, b \in \mathscr{F}$, then $a(-b) = -ab$, $(-a)b = -ab$, and $(-a)(-b) = ab$.

PROOF OF B_{11}. Let $a, b \in \mathscr{F}$. We shall establish the first equality only. Then

$$\begin{aligned} ab + a(-b) &= a(b + (-b)) \\ &= a \cdot 0 \\ &= 0. \end{aligned}$$

In addition, $a(-b) + ab = ab + a(-b) = 0$. Thus by B_2, $-ab = a(-b)$.

For $a, b, c, d \in \mathscr{F}$, define $abc = (ab)c$ and $abcd = (abc)d$. Then $abc = a(bc)$.

B_{12}. If $a, b, c, d \in \mathscr{F}$, $a(bc)d = (ab)(cd)$ and $a(bcd) = (ab)(cd)$.

PROOF OF B_{12}. Let $a, b, c, d \in \mathscr{F}$. We shall establish the first equality only. Then

$$\begin{aligned} a(bc)d &= (a(bc))d \\ &= a((bc)d) \\ &= a(b(cd)) \\ &= (ab)(cd). \end{aligned}$$

B_{13}. If $a, b, c, d, e \in \mathscr{F}$, $(abc)(de) = a(bcde)$.

This is stated without proof.

QUATERNIONS APPENDIX C

The system of quaternions, representing an example of a division ring which is not a field, is used in Chapter 7 to construct an instance of a desarguesian projective plane which is not pappian. We shall present here a brief discussion of this system. Any statements not proved are left as exercises.

Let $(\mathscr{R}, +, \cdot)$ be the field of real numbers. We shall write $ab = a \cdot b$ for a $b \in \mathscr{R}$. Define

$$\mathscr{D} = \{(a, b, c, d)\colon a, b, c, d \in \mathscr{R}\}.$$

The elements of $\mathscr{D}$ will be called *quaternions*.

Define $+$, a binary operation on $\mathscr{D} \times \mathscr{D}$ to $\mathscr{D}$ called *addition*, by

$$(a, b, c, d) + (a', b', c', d') = (a + a', b + b', c + c', d + d')$$

for $a, b, c, d, a', b', c', d' \in \mathscr{R}$. (Note that the symbol $+$ appearing on the left-hand side of the equality is the addition for quaternions being defined and $+$ on the right-hand side is addition for real numbers. It is customary in this situation to let the symbol $+$ have these two meanings. The meaning intended in a specific usage should be evident from the context.)

Define $\cdot$, a binary operation on $\mathscr{D} \times \mathscr{D}$ to $\mathscr{D}$ called *multiplication*, by

$$(a, b, c, d) \cdot (a', b', c', d') = (aa' - bb' - cc' - dd', ab' + ba' + cd' - dc', ac' - bd' + ca' + db', ad' + bc' - cb' + da')$$

for $a, b, c, d, a', b', c', d' \in \mathscr{R}$. (Note that the symbol $\cdot$ appearing on the left-hand side of the equality is the multiplication being defined for quaternions.)

C₁. For $\mathbf{x}, \mathbf{y}, \mathbf{z} \in \mathscr{Q}$, $\mathbf{x} + (\mathbf{y} + \mathbf{z}) = (\mathbf{x} + \mathbf{y}) + \mathbf{z}$.

PROOF OF C_1. Let $\mathbf{x}, \mathbf{y}, \mathbf{z} \in \mathscr{Q}$. Then $\mathbf{x} = (a, b, c, d)$ $\mathbf{y} = (a', b', c', d')$, and $\mathbf{z} = (a'', b'', c'', d'')$ for some $a, b, c, d, a', b', c', d', a'', b'', c'', d'' \in \mathscr{R}$. Now

$$\begin{aligned}
\mathbf{x} + (\mathbf{y} + \mathbf{z}) &= (a, b, c, d) + ((a', b', c', d') + (a'', b'', c'', d'')) \\
&= (a, b, c, d) + (a' + a'', b' + b'', c' + c'', d' + d'') \\
&= (a + (a' + a''), b + (b' + b''), c + (c' + c''), d + (d' + d'')) \\
&= ((a + a') + a'', (b + b') + b'', (c + c') + c'', (d + d') + d'') \\
&= (a + a', b + b', c + c', d + d') + (a'', b'', c'', d'') \\
&= ((a, b, c, d) + (a', b', c', d')) + (a'', b'', c'', d'') \\
&= (\mathbf{x} + \mathbf{y}) + \mathbf{z}.
\end{aligned}$$

Now we can define for $\mathbf{w}, \mathbf{x}, \mathbf{y}, \mathbf{z} \in \mathscr{Q}$,

$$\mathbf{x} + \mathbf{y} + \mathbf{z} = (\mathbf{x} + \mathbf{y}) + \mathbf{z}$$

and

$$\mathbf{x} + \mathbf{y} + \mathbf{z} + \mathbf{w} = (\mathbf{x} + \mathbf{y} + \mathbf{z}) + \mathbf{w}.$$

The mechanics of computing with quaternions can be simplified considerably by the introduction of symbols for a few specific quaternions and the adoption of some conventions. Define $\mathbf{i} = (0, 1, 0, 0)$, $\mathbf{j} = (0, 0, 1, 0)$, and $\mathbf{k} = (0, 0, 0, 1)$. We shall write $a = (a, 0, 0, 0)$ for $a \in \mathscr{R}$ and $\mathbf{xy} = \mathbf{x} \cdot \mathbf{y}$ for $\mathbf{x}, \mathbf{y} \in \mathscr{Q}$. Then we have the following.

C₂. For $a, b, c, d \in \mathscr{R}$,

(a) $b\mathbf{i} = (0, b, 0, 0)$;
(b) $c\mathbf{j} = (0, 0, c, 0)$;
(c) $d\mathbf{k} = (0, 0, 0, d)$;
(d) $a + b\mathbf{i} + c\mathbf{j} + d\mathbf{k} = (a, b, c, d)$.

PROOF OF C_2. Let $a, b, c, d \in \mathscr{R}$. For part (a),

$$\begin{aligned} b\mathbf{i} &= b \cdot \mathbf{i} \\ &= (b, 0, 0, 0) \cdot (0, 1, 0, 0) \\ &= (b \cdot 0 - 0 \cdot 1 - 0 \cdot 0 - 0 \cdot 0, b \cdot 1 + 0 \cdot 0 + 0 \cdot 0 - 0 \cdot 0, \\ &\qquad b \cdot 0 - 0 \cdot 0 + 0 \cdot 0 + 0 \cdot 1, b \cdot 0 + 0 \cdot 0 - 0 \cdot 1 + 0 \cdot 0) \\ &= (0, b, 0, 0). \end{aligned}$$

For part (d),

$$\begin{aligned} a + b\mathbf{i} + c\mathbf{j} + d\mathbf{k} &= ((a + b\mathbf{i}) + c\mathbf{j}) + d\mathbf{k} \\ &= (((a, 0, 0, 0) + (0, b, 0, 0)) + (0, 0, c, 0)) + (0, 0, 0, d) \\ &= ((a, b, 0, 0) + (0, 0, c, 0)) + (0, 0, 0, d) \\ &= (a, b, c, 0) + (0, 0, 0, d) \\ &= (a, b, c, d). \end{aligned}$$

We shall write $\mathbf{x}^2 = \mathbf{x} \cdot \mathbf{x}$ and $-\mathbf{x} = (-1)\mathbf{x}$ for $\mathbf{x} \in \mathscr{D}$.

C_3. The quaternions $\mathbf{i}, \mathbf{j}, \mathbf{k}$ satisfy multiplication according to the following table.

	i	**j**	**k**
i	-1	$\mathbf{k}$	$-\mathbf{j}$
j	$-\mathbf{k}$	-1	$\mathbf{i}$
k	$\mathbf{j}$	$-\mathbf{i}$	-1

PROOF OF C_3. We shall establish the validity of only two entries. We have

$$\begin{aligned} \mathbf{i}^2 &= (0, 1, 0, 0) \cdot (0, 1, 0, 0) \\ &= (0 \cdot 0 - 1 \cdot 1 - 0 \cdot 0 - 0 \cdot 0, 0 \cdot 1 + 1 \cdot 0 + 0 \cdot 0 - 0 \cdot 0, \\ &\qquad 0 \cdot 0 - 1 \cdot 0 + 0 \cdot 0 + 0 \cdot 1, 0 \cdot 0 + 1 \cdot 0 - 0 \cdot 1 + 0 \cdot 0) \\ &= (-1, 0, 0, 0) \\ &= -1 \end{aligned}$$

and

$$\begin{aligned}\mathbf{kj} &= (0, 0, 0, 1) \cdot (0, 0, 1, 0)\\ &= (0 \cdot 0 - 0 \cdot 0 - 0 \cdot 1 - 1 \cdot 0, 0 \cdot 0 + 0 \cdot 0 + 0 \cdot 0 - 1 \cdot 1,\\ &\qquad 0 \cdot 1 - 0 \cdot 0 + 0 \cdot 0 + 1 \cdot 0, 0 \cdot 0 + 0 \cdot 1 - 0 \cdot 0 + 1 \cdot 0)\\ &= (0, -1, 0, 0)\\ &= (-1)\mathbf{i}\\ &= -\mathbf{i}.\end{aligned}$$

The above definitions and properties are summarized in the next statement which gives the addition and multiplication of quaternions an algebraic representation resembling expressions from college algebra.

C_4. Let $a, b, c, d, a', b', c', d' \in \mathscr{R}$. Then

(a) $(a + b\mathbf{i} + c\mathbf{j} + d\mathbf{k}) + (a' + b'\mathbf{i} + c'\mathbf{j} + d'\mathbf{k})$
$\qquad = (a + a') + (b + b')\mathbf{i} + (c + c')\mathbf{j} + (d + d')\mathbf{k};$

(b) $(a + b\mathbf{i} + c\mathbf{j} + d\mathbf{k}) \cdot (a' + b'\mathbf{i} + c'\mathbf{j} + d'\mathbf{k})$
$\qquad = (aa' - bb' - cc' - dd') + (ab' + ba' + cd' - dc')\mathbf{i}$
$\qquad\quad + (ac' - bd' + ca' + db')\mathbf{j} + (ad' + bc' - cb' + da')\mathbf{k}.$

C_5. The system $(\mathscr{D}, +, \cdot)$ is a division ring but not a field.

PROOF OF C_5. We shall refer to Appendix B for the definition of a division ring, which includes the eight axioms to be proved for $(\mathscr{D}, +, \cdot)$. The associativity of $+$, the first axiom, is Statement C_1. We shall consider only some of the remaining axioms.

For the second axiom, let $a, b, c, d \in \mathscr{R}$. Then

$$\begin{aligned}(a + b\mathbf{i} + c\mathbf{j} + d\mathbf{k}) + 0 &= (a + b\mathbf{i} + c\mathbf{j} + d\mathbf{k}) + (0 + 0\mathbf{i} + 0\mathbf{j} + 0\mathbf{k})\\ &= a + b\mathbf{i} + c\mathbf{j} + d\mathbf{k}.\end{aligned}$$

Similarly, $0 + (a + b\mathbf{i} + c\mathbf{j} + d\mathbf{k}) = a + b\mathbf{i} + c\mathbf{j} + d\mathbf{k}$.

For Axiom 6, we have 1 the unique element of $\mathscr{D}$ such that $\mathbf{x} \cdot 1 = \mathbf{x}$ and $1 \cdot \mathbf{x} = \mathbf{x}$ for $\mathbf{x} \in \mathscr{D}$.

For Axiom 7, let $\mathbf{x} \in \mathscr{D}$ such that $\mathbf{x} \neq 0$. Then $\mathbf{x} = a + b\mathbf{i} + c\mathbf{j} + d\mathbf{k}$ for some $a, b, c, d \in \mathscr{R}$ such that not all $a = 0$, $b = 0$, $c = 0$, and $d = 0$ hold. Now $a^2 + b^2 + c^2 + d^2 \neq 0$. Let $e = a^2 + b^2 + c^2 + d^2$. Then

$$(a + b\mathbf{i} + c\mathbf{j} + d\mathbf{k}) \cdot \left(\frac{a}{e} - \frac{b}{e}\mathbf{i} - \frac{c}{e}\mathbf{j} - \frac{d}{e}\mathbf{k}\right)$$

$$= \frac{a^2 + b^2 + c^2 + d^2}{e} + \frac{-ab + ba - cd + dc}{e}\mathbf{i}$$

$$+ \frac{-ac + bd + ca - db}{e}\mathbf{j} + \frac{-ad - bc + cb + da}{e}\mathbf{k}$$

$$= 1.$$

Similarly,

$$\left(\frac{a}{e} - \frac{b}{e}\mathbf{i} - \frac{c}{e}\mathbf{j} - \frac{d}{e}\mathbf{k}\right) \cdot (a + b\mathbf{i} + c\mathbf{j} + d\mathbf{k}) = 1.$$

Finally, to show that $(\mathscr{D}, +, \cdot)$ is not a field, consider

$$\mathbf{i} \cdot \mathbf{j} = \mathbf{k} = (0, 0, 0, 1) \neq (0, 0, 0\ -1) = -\mathbf{k} = \mathbf{j} \cdot \mathbf{i}.$$

REFERENCES

1. Albert, A. A., *On non-associative division algebras, Trans. Amer. Math. Soc.*, **72** (1952), 296–309.
2. Albert, A. A., and R. Sandler, "An Introduction to Finite Projective Planes." Holt, New York, 1968.
3. The American Mathematical Monthly, Number 4 of the Slaught Memorial Papers, *Contributions to Geometry, Math. Assoc. Amer.*, **62** (1955).
4. Artin, E., "Geometric Algebra," Wiley (Interscience), New York, 1957.
5. Baer, R. "Linear Algebra and Projective Geometry." Academic Press, New York, 1952.
6. Blumenthal, L., "A Modern View of Geometry." Freeman, San Francisco, 1961.
7. Bruck, R., and H. Ryser, *The nonexistence of certain finite projective planes, Can. J. Math*, **1** (1949), 88–93.
8. Coxeter, H. S. M., "The Real Projective Plane." McGraw-Hill, New York, 1959.
9. Eves, H., and C. V. Newsom, "An Introduction to the Foundations and Fundamental Concepts of Mathematics." Holt, New York, 1958.
10. Halmos, P. R., "Naive Set Theory," Van Nostrand Reinhold, Princeton, New Jersey, 1960.
11. Herstein, I. N., "Topics in Algebra," Ginn (Blaisdell), Boston, 1964.
12. Hilbert D., "The Foundations of Geometry" (translated by Leo Unger, revised by Paul Bernays), 10th ed., Open Court, LaSalle, Illinois, 1971.
13. Kershner, R. B., and L. R. Wilcox, "The Anatomy of Mathematics." Ronald Press, New York, 1950.
14. Menger, K., *Self-dual postulates in projective geometry, Amer. Math. Monthly*, **55** (1948), 195.
15. Prenowitz, W., and M. Jordan, "Basic Concepts of Geometry," Ginn (Blaisdell), Boston, 1965.

16. Seidenberg, A., "Lectures in Projective Geometry." Van Nostrand Reinhold, Princeton, New Jersey, 1962.
17. Tarry, G., *Le probleme des 36 officiers*, *Camptes Revdn. Ass. Fr. pour l'Avancement de Science Naturel*, **1** (1900), 122–123; **2** (1901), 170–203.
18. Tuller, A., "A Modern Introduction to Geometries." Van Nostrand Reinhold, Princeton, New Jersey, 1967.
19. Veblen, O., and J. W. Young, "Projective Geometry." 2 vols. Ginn, Boston, 1910 and 1918.
20. Wilder, R. L., "Introduction to the Foundations of Mathematics." Wiley, New York, 1952.

INDEX OF SPECIAL SYMBOLS

SUBJECT INDEX